AI in the Apiary ©
HONEYBEE DISEASES

Paddy G Walker

NB

Northern Bee Books

AI in the Apiary © HONEYBEE DISEASES
Copyright © Paddy G Walker

Published 2024 by
Northern Bee Books,
Scout Bottom Farm,
Mytholmroyd,
West Yorkshire
HX7 5JS (UK)
Tel: 01422 882751
Fax: 01422 886157
www.northernbeebooks.co.uk

ISBN 978-1-914934-78-0

Design and artwork DM Design and Print

Cover Images:
a. Courtesy USDA ARS (Phil Franks)

b. Deformed Wing Virus
   Courtesy Rcsb.org
   5g52
   Crystallographic Structure Of Full Particle Of Deformed Wing Virus
   Skubnik, K., Novacek, J., Fuzik, T., Pridal, A., Paxton, R., Plevka, P.
   (2017) Proc Natl Acad Sci U S A 114: 3210-3215

c. Chronic Bee Paralysis Virus
   Courtesy Animal And Plant Health Agency © Crown Copyright

Foreword

AI in the Apiary

After homo sapiens, apis mellifera, or a variation thereof, is the most studied (kingdom) animalia in the world. A huge amount of information exists relating to honeybees, going back many years into the 19th century. The challenge is how to access this massive amount of information in a meaningful manner.

Using a regular search engine provides far too much information to easily make sense of in a practical manner. If the topic, such as honeybee diseases is studied, not only is it hard to access in an effective manner, it is also almost impossible to access consistent sets of data relating to any diseases. For example, if deformed wing virus (DWV) is studied, a standard internet search produces in excess of 250,000 responses. This level of response if not useful.

Artificial intelligence (AI) takes a very different approach, which is very complete, can be very consistent (but not always) and produces data sets which are way beyond the capability of traditional research methods. AI is a large language model (LLM) which allows for a comprehensive and complex assessment of data sitting in more than a trillion 'bits' of information.

Getting a reasonable response then becomes very possible. The next step is to get a response to a more complex question and so to produce a (reasonably) uniformly structured output across a wide range of different types of honeybee diseases.

Initially, the scope of this book requires a clear definition. To a larger extent, the coverage is many but not all, honeybee diseases, typically (not exclusively) viral, fungal, bacterial and parasitic. Also covered are colony collapse disorder (CCD), addled brood and chilled brood.

This presents a simple definition of scope. The next challenge is to determine what information makes for an informative set which is very useful to a beekeeper. This was established as follows, starting with the 7 higher level classification:

Classification

- Kingdom/Realm
- Phylum
- Class
- Order
- Family
- Genus
- Species

Biology

Transmission

Symptoms

How to Identify the Disease

Impact on the Colonies

Advice on Mitigation

The requirement from an AI LLM is to enable the AI to learn what makes the necessary output that can consistently present, to a larger degree, content which both accommodates the scope of this book and the clear structure of the developed content.

The learning by the AI LLM was encouraged by repeatedly developing and trialling a question set until the output was both useful and could be replicated. All responses need some changes but more in terms of structure and presentation than content. In some cases the information requested was either not available or a long way from being complete or completable. An example would be varroa destructor virus (VDV). There are at least 10 variants of this virus, known VDV1, VDV2, etc. more than half have only been reported in 2023 (from China) and have no additional data other than to indicate a variant exists. Hence this book only covers VDV1 (which is now classified as DWV-B) and VDV2. Additionally, it should be noted that mucor spp. and penicillium spp. for example, are collective disease names, which indicate more than a single variant. Mucor spp. has some 47 variants, for example. These are reported therefore collectively rather than in detail.

Finally, the limited number of illustrations are indicative of a dearth of honeybee disease images, whether copyright or not. This included a very wide trawl covering Canada, USA (USAD) Europe and Australia. *Also due to the nature of AI LLMs, no possibility of attribution exists, so the author of this book has not made any text acknowledgements. That is not for want of trying.* Images are fully acknowledged.

The biological models are 3d and are courtesy of rcsb.org. Using the 4 digit model identifier in the search facility of the site will allow the reader to both rotate the models and expand fully to 'see inside'.

At this point it is valid to look at the classification of diseases in terms of viral, bacterial, fungal and parasitic, along with more detail.

Honeybees, despite their vital role in pollination and honey production, face various threats, including a range of parasitic diseases. Understanding these diseases and their classification is crucial for beekeepers to protect their colonies. Here are some examples of parasitic diseases affecting honeybees, categorised based on the disease type:

1. Mite Infestations:

Varroa mite (Varroa destructor): The most significant parasite, feeding on bee hemolymph and transmitting viruses like Deformed Wing Virus. This widespread mite can lead to colony collapse.

Acarine disease (Acarapis woodi): Infests bee tracheas, causing respiratory distress and death. This disease was once widespread but is less common today.

2. Other Parasitic Infestations:

Tropilaelaps mite (Tropilaelaps spp.): Found primarily in tropical regions, this mite feeds on bee hemolymph and can compete with Varroa mites.

Braula fly (Braula coeca): This wingless fly lives on adult bees and feeds on their hemolymph. While not as damaging as other parasites, it can contribute to colony stress.

3. Fungal Diseases:

Chalkbrood (Ascosphaera apis): Caused by a fungus, this disease mummifies larvae, with white spores visible on their bodies. While not as aggressive as some other diseases, it can weaken colonies.

Nosemosis (Nosema apis): Caused by a single-celled organism, Nosema apis, impacting the bee digestive system and weakening the colony. This is a common and persistent disease.

4. Viral Diseases (often transmitted by parasites):

Deformed Wing Virus (DWV): Highly prevalent, transmitted by the Varroa mite, this virus causes deformed wings and hinders bee flight ability.

Acute Bee Paralysis Virus (ABPV): Also associated with Varroa mite infestations, ABPV can cause rapid colony decline and paralysis in adult bees.

5. Viral Diseases (often transmitted by parasites):

Deformed Wing Virus (DWV): Highly prevalent, transmitted by the Varroa mite, this virus causes deformed wings and hinders bee flight ability.

Acute Bee Paralysis Virus (ABPV): Also associated with Varroa mite infestations, ABPV can cause rapid colony decline and paralysis in adult bees.

Note: This list is not exhaustive. Early detection and intervention are crucial for controlling parasitic diseases and protecting honeybee health. Beekeepers should be vigilant, perform regular hive inspections and implement appropriate treatment strategies when necessary.

Honeybee diseases can be classified in several ways, offering different perspectives on their origins and potential interventions. Here are some common classification methods:

By causative agent:

Parasitic: Caused by mites, beetles, or larvae that feed on bees or their brood. Examples include Varroa destructor mites and small hive beetle infestation.

Bacterial: Caused by bacterial infections of the brood or adult bees. Examples include American foulbrood (AFB) and European foulbrood (EFB).

Fungal: Caused by fungal infections that grow on brood or other bee products. Examples include chalkbrood and stonebrood.

Viral: Caused by viruses that infect bees and disrupt their functions. Examples include chronic bee paralysis virus (CBPV) and deformed wing virus (DWV).

By affected individuals:

Brood diseases: Primarily affect larvae or pupae (developing bees). Examples include AFB, EFB, chalkbrood, and stonebrood.

Adult bee diseases: Primarily affect worker bees, drones, or queens. Examples include Nosema disease, acarine disease, and foulbrood.

By environmental factors:

Stress-related: May be triggered by malnutrition, pesticide exposure, poor weather conditions, or colony mismanagement. Examples include Colony Collapse Disorder (CCD) and dysentery.

Temperature-related: May become more prevalent under extreme temperatures or fluctuations. Examples of temperature-sensitive fungal diseases like chalkbrood and stonebrood.

By economic impact:

Major diseases: Cause significant colony losses and economic damage to beekeepers. Examples include AFB, EFB, and Varroa infestation.

Minor diseases: Typically have less severe economic impact but can still harm colony health. Examples include Nosema disease and acarine disease.

Additional classification systems:

By transmissibility: How easily the disease can spread within and between colonies.

By symptoms: Observable signs and behaviours of infected bees.

By treatment options: Availability and effectiveness of control measures.

Understanding these different classification schemes helps beekeepers, researchers, and policymakers effectively manage honeybee health and combat the threats posed by various diseases.

The Classification of Viruses:
Unveiling the World of Viral Diseases

A Comprehensive Guide

Viruses, the microscopic masters of hijacking cellular machinery, come in a dizzying array of shapes, sizes, and genetic compositions. Classifying these infectious agents is crucial for understanding their origins, evolution, and potential threats to human and animal health. This guide delves into the fascinating world of viral classification, unveiling the different systems and criteria used to organise these diverse entities.

1. Unveiling the Tools of the Trade:
Before diving into the classification systems, let's equip ourselves with some key tools:

Morphology: The virus's physical shape, like icosahedral (spherical with 20 faces), helical (cylindrical), or enveloped (surrounded by a membrane).

Genome: The genetic material, either DNA or RNA, single-stranded (ss) or double-stranded (ds).

Replication Strategy: How the virus replicates inside host cells, involving processes like reverse transcription or RNA-dependent RNA polymerase.

2. The Hierarchical Ladder:
Viral classification follows a hierarchical system, similar to how we categorise living organisms. Here's how it works:

Realm: The broadest category, based on fundamental features like genome type (DNA or RNA).

Phylum: Defined by the type of nucleic acid (DNA or RNA) and its strandedness (single-stranded or double-stranded).

Class: Determined by the presence or absence of an envelope and the replication strategy.

Order: Characterised by specific features like virion morphology and capsid protein arrangement.

Family: Groups viruses with similar genomes, replication strategies, and protein structures.

Genus: Categorises viruses within a family based on more detailed genetic and antigenic relationships.

Species: The most specific level, representing viruses with highly similar genomes and biological properties.

3. The Guiding Principles:
Several key principles guide viral classification:

Phylogenetic Relationships: Ideally, the classification reflects the evolutionary history of viruses, grouping related ones together.

Shared Properties: Viruses are grouped based on common characteristics like genome type, morphology, and replication strategy.

Disease Association: Viruses causing similar diseases may be placed close to each other, aiding in diagnosis and treatment.

4. The Authority in Charge:
The International Committee on Taxonomy of Viruses (ICTV) is the official body responsible for viral classification. This global organisation regularly reviews and updates the viral taxonomy, ensuring consistency and accuracy.

5. Beyond the Basics:
Viral classification is an ongoing endeavour, constantly evolving as new viruses are discovered and our understanding of their diversity deepens. Additional classification systems are sometimes used for specific purposes, such as grouping viruses based on their host range or the diseases they cause.

Understanding the classification of viruses is essential for:
Developing vaccines and antiviral drugs: Targeting specific viral groups or families can lead to more effective treatments.

Tracking virus outbreaks: Classifying newly discovered viruses helps predict their potential spread and impact.

Conserving biodiversity: Studying viral diversity contributes to understanding the delicate balance of ecosystems.

By delving into the world of viral classification, we gain a deeper appreciation for the incredible variety and complexity of these tiny life forms. This knowledge empowers us to better understand, manage, and potentially even prevent future viral threats.

Unveiling the World of Bacterial Diseases

A Comprehensive Guide to Classification

Bacteria, despite their single-celled simplicity, are responsible for a vast array of diseases in humans, animals, and plants. Navigating this diverse landscape requires a robust classification system, allowing us to understand, diagnose, and treat these infections effectively. This comprehensive guide delves into the intricate world of bacterial disease classification, exploring the various systems used to categorise these microscopic culprits.

1. Unveiling the Hierarchy: Ranks and Levels
Bacterial disease classification, like any biological classification system, follows a hierarchical structure. Bacterial pathogens are grouped into progressively broader categories based on shared characteristics. Here's a breakdown of the main ranks:

Kingdom: All bacteria belong to the Bacteria kingdom within the Domain Bacteria.

Phylum: The kingdom is further divided into phyla based on cell wall characteristics and metabolic pathways. Some major phyla include Firmicutes, Proteobacteria, Bacteroidetes, and Actinobacteria.

Class: Within each phylum, classes are defined by specific cell wall features and metabolic capabilities. For example, the Firmicutes phylum contains the classes Clostridia (spore-forming) and Bacilli (non-spore-forming).

Order: Orders further refine the classification based on more detailed morphological and biochemical characteristics. For example, the Bacilli class includes the order Lactobacillales (lactic acid bacteria) and the order Bacillales (rod-shaped bacteria).

Family: Families group bacteria with even closer shared characteristics, including genome organisation and phylogenetic relationships. Some well-known families include Enterobacteriaceae (intestinal bacteria) and Staphylococcaceae (staphylococci).

Genus: The genus level represents a group of closely related bacterial species with highly similar genetic and phenotypic characteristics. For example, the Enterobacteriaceae family includes the genera Escherichia (E. coli), Salmonella, and Shigella.

Species: Finally, the species level is the most specific grouping, encompassing bacteria with nearly identical genomes and typically causing the same disease. Examples include Escherichia coli O157:H7 (associated with foodborne illness) and Staphylococcus aureus (responsible for skin infections and pneumonia).

2. Demystifying the Criteria: What Makes a Bacterial Disease Unique?
So, what specific characteristics do researchers use to classify bacterial diseases? Here are some key considerations:

Bacterial species: The specific type of bacteria involved in the disease, identified by its genus and species.

Host range: The types of organisms the bacteria can infect, such as humans, animals, or plants.

Mode of transmission: How the bacteria spreads from one host to another, such as through direct contact, airborne transmission, or contaminated food or water.

Pathogenesis: The mechanisms by which the bacteria cause disease, including toxin production, tissue invasion, and immune system response.

Clinical manifestations: The specific symptoms and signs associated with the disease, such as fever, cough, diarrhea, or skin lesions.

Antibiotic susceptibility: The effectiveness of different antibiotics against the bacteria.

3. Beyond the Ranks: Alternative Classification Systems
While the hierarchical system based on bacterial taxonomy is widely used, other classification approaches exist, focusing on specific aspects of bacterial diseases. For example, some systems group diseases based on the anatomical site of infection (e.g., respiratory infections, gastrointestinal infections) or the severity of the disease (e.g., mild, moderate, severe). Additionally, functional classification systems categorise diseases based on the toxins produced by the bacteria or the cellular processes they target.

4. The Authority: The International Committee on Systematics of Bacteria (ICSSB)
The ICSSB is the official body responsible for bacterial taxonomy and nomenclature. This international committee of experts regularly reviews and updates the bacterial taxonomy based on new discoveries and advancements in microbiological research. Their website provides a comprehensive and up-to-date resource on bacterial classification, including detailed descriptions of each taxonomic rank and taxon.

5. A Dynamic Landscape: The Ever-Evolving World of Bacteria
Bacterial disease classification is not static. As new bacterial species are discovered and our understanding of existing ones deepens, the taxonomic landscape constantly evolves. New ranks and taxa may be added, existing ones may be revised, and classification criteria may be refined. This dynamic nature reflects the incredible diversity and adaptability of bacteria, highlighting the need for continuous research and refinement in our understanding of these microscopic pathogens.

Unveiling the Fungal Kingdom:

A Comprehensive Guide to Fungal Disease Classification

Fungi, despite their seemingly simple nature, boast a vast and diverse kingdom capable of both symbiosis and pathogenesis. Classifying their disease-causing members is crucial for proper diagnosis, treatment, and prevention. This guide delves into the intricate world of fungal disease classification, exploring the approaches used to categorise these multifaceted organisms.

1. Unveiling the Hierarchy: Ranks and Levels

Similar to other biological classifications, fungal disease categorisation follows a hierarchical structure. Fungal pathogens are grouped into progressively broader categories based on shared characteristics. Here's a breakdown of the main ranks:

Kingdom: All fungi belong to the Fungi kingdom within the Domain Eukaryota.

Phylum: The kingdom is further divided into phyla based on morphological features like cell wall composition and reproductive structures. Major phyla include Ascomycota, Basidiomycota, and Sygomycota.

Class: Within each phylum, classes are defined by specific morphologies and life cycles. For example, the Ascomycota phylum contains the classes Saccharomycetes (yeasts) and Sordariomycetes (filamentous fungi).

Order: Orders refine the classification based on more detailed morphological and ecological characteristics. An example from the Sordariomycetes class includes the order Hypocreales (contains fungi causing plant diseases and some human infections).

Family: Families group fungi with closer shared characteristics, including genome organisation and phylogenetic relationships. Some well-known families include Trichocomaceae (dermatophytes causing skin infections) and Candidaceae (yeasts contributing to opportunistic infections).

Genus: The genus level represents a group of closely related fungal species with highly similar genetic and phenotypic characteristics. Examples include Trichophyton (skin infections) and Candida (vaginal yeast infections).

Species: Finally, the species level is the most specific, encompassing fungi with nearly identical genomes and typically causing the same disease. Examples include Trichophyton rubrum (ringworm) and Candida albicans (vaginal yeast infections).

2. Demystifying the Criteria: What Makes a Fungal Disease Unique?

So, what specific characteristics do researchers use to classify fungal diseases? Here are some key considerations:

Fungal species: The specific type of fungus involved in the disease, identified by its genus and species.

Host range: The types of organisms the fungus can infect, such as humans, animals, or plants.

Mode of transmission: How the fungus spreads from one host to another, such as through direct contact, airborne spores, or contaminated soil.

Pathogenesis: The mechanisms by which the fungus causes disease, including tissue invasion, toxin production, and immune system response.

Clinical manifestations: The specific symptoms and signs associated with the disease, such as skin lesions, respiratory problems, or opportunistic infections.

Antifungal susceptibility: The effectiveness of different antifungal medications against the fungus.

3. Beyond the Ranks: Alternative Classification Systems

While the hierarchical system based on fungal taxonomy is widely used, other classification approaches exist, focusing on specific aspects of fungal diseases. For example, some systems group diseases based on the anatomical site of infection (e.g., dermatomycoses, pneumomycoses) or the severity of the disease (e.g., superficial, subcutaneous, systemic). Additionally, functional classification systems categorise diseases based on the toxins produced by the fungi or the cellular processes they target.

4. The Authority: The International Commission on the Taxonomy of Fungi (ICTF)

The ICTF is the official body responsible for fungal taxonomy and nomenclature. This international committee of experts regularly reviews and updates the fungal taxonomy based on new discoveries and advancements in mycological research. Their website provides a comprehensive and up-to-date resource on fungal classification, including detailed descriptions of each taxonomic rank and taxon.

5. A Dynamic Landscape: The Ever-Evolving World of Fungi

Fungal disease classification is not static. As new fungal species are discovered and our understanding of existing ones deepens, the taxonomic landscape constantly evolves. New ranks and taxa may be added, existing ones may be revised, and classification criteria may be refined. This dynamic nature reflects the incredible diversity and adaptability of fungi, highlighting the need for continuous research and refinement in our understanding of these fascinating organisms.

Unveiling the World of Parasitic Diseases

A Comprehensive Guide to Parasitic Disease Classification

Parasites, despite their often-overlooked existence, play a significant role in shaping the health of living organisms, including humans. Understanding the diverse tapestry of parasites and the diseases they cause is crucial for effective diagnosis, treatment, and prevention. This guide delves into the intricate world of parasitic disease classification, exploring the systems used to categorise these hidden invaders.

1. Unveiling the Hierarchy: Ranks and Levels

Similar to other biological classifications, parasite disease categorisation follows a hierarchical structure. Parasites are grouped into progressively broader categories based on shared characteristics. Here's a breakdown of the main ranks:

Kingdom: All parasites belong to various kingdoms within the Eukaryota domain, depending on their specific cellular features. This includes Animalia (roundworms, flatworms, arthropods), Plantae (fungal-like parasites), and Chromista (unicellular protists).

Class: Classes further refine the classification based on more detailed morphological features. Some well-known classes include Trematoda (flukes) and Cestoda (tapeworms) within the Platyhelminthes phylum.

Order: Orders further refine the classification based on specific life cycles and developmental stages. An example from the Nematoda phylum includes the order Ascaridida (contains roundworms causing intestinal infections).

Family: Families group parasites with closer shared characteristics, including internal anatomy, host range, and transmission mechanisms. Some well-known families include Taeniidae (tapeworms) and Strongylidae (hookworms).

Genus: The genus level represents a group of closely related parasite species with highly similar genetic and phenotypic characteristics. Examples include Taenia (tapeworms) and Ascaris (roundworms).

Species: Finally, the species level is the most specific, encompassing parasites with nearly identical genomes and typically causing the same disease. Examples include Taenia solium (pork tapeworm) and Ascaris lumbricoides (common roundworm).

2. Demystifying the Criteria: What Makes a Parasitic Disease Unique?

So, what specific characteristics do researchers use to classify parasitic diseases? Here are some key considerations:

Parasite species: The specific type of parasite involved in the disease, identified by its genus and species.

Host range: The types of organisms the parasite can infect, such as humans, animals, or plants.

Life cycle: The stages of development the parasite undergoes within its host and intermediate hosts, if any.

Mode of transmission: How the parasite spreads from one host to another, such as through direct contact, contaminated food or water, or insect vectors.

Pathogenesis: The mechanisms by which the parasite causes disease, including tissue invasion, nutrient depletion, toxin production, and immune system response.

Clinical manifestations: The specific symptoms and signs associated with the disease, such as abdominal pain, diarrhea, anemia, or skin lesions.

Antifungal or antiparasitic susceptibility: The effectiveness of different medications against the parasite.

3. Beyond the Ranks: Alternative Classification Systems

While the hierarchical system based on parasite taxonomy is widely used, other classification approaches exist, focusing on specific aspects of parasitic diseases. For example, some systems group diseases based on the geographical distribution of the parasite or the organ system affected (e.g., intestinal parasites, blood parasites). Additionally, functional classification systems categorise diseases based on the type of immune response they elicit.

4. The Authority:

The International Commission on Soological Nomenclature (ICSN)
The ICSN is the official body responsible for animal taxonomy and nomenclature. This international commission periodically reviews and updates animal classifications, which includes many parasitic groups. Additionally, various organisations and research groups contribute to specific parasite classifications within their areas of expertise.

5. A Dynamic Landscape:

The Ever-Evolving World of Parasites
Parasitic disease classification is not static. As new parasite species are discovered and our understanding of existing ones deepens, the taxonomic landscape constantly evolves. New ranks and taxa may be added, existing ones may be revised, and classification criteria may be refined. This dynamic nature reflects the incredible diversity and adaptability of parasites, highlighting the need for continuous research and refinement in our understanding of these fascinating, yet potentially harmful, organisms.

HONEYBEE DISEASES

1. ACARAPISOSIS

Acarapisosis is also known as Isle of Wight Disease (IOWD)

Acarapisosis, also referred to as acariosis or acarine disease, is a disease of honeybees caused by the parasitic mite Acarapis woodi. The disease is classified as follows:

▶ Kingdom: Animalia
▶ Phylum: Arthropoda
▶ Class: Arachnida
▶ Order: Trombidiformes
▶ Family: Tarsonemidae
▶ Genus: Acarapis
▶ Species: Acarapis woodi

Biology

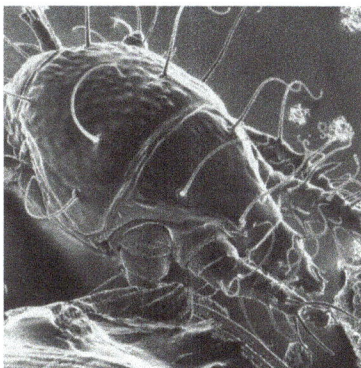

Tracheal mite (*Acarapis woodi*) Scanning Electron Microscope image of Tracheal mite (*Acarapis woodi*)

Courtesy the Animal and Plant Health Agency © Crown Copyright

Acarapis woodi is a microscopic mite that lives in the trachea (breathing tubes) of honeybees. It is about 0.15-0.2 mm long and has a cigar-shaped body. The mite has four pairs of legs and two pairs of palps (mouthparts).

The life cycle of Acarapis woodi is completed in 7-10 days. The female mite lays her eggs in the trachea of a honeybee. The eggs hatch into larvae, which develop into nymphs and then into adult mites. The adult mites mate and the female mites lay more eggs.

Transmission

Acarapis woodi is transmitted from bee to bee through contact. The mites can also be transmitted through contaminated hive equipment.

Symptoms

▶ The symptoms of acarapisosis include:
▶ Shortening of the lifespan of adult bees
▶ Reduced honey production
▶ Increased swarming
▶ Increased susceptibility to other diseases
▶ Difficulty flying

- Crawling on the ground in front of the hive
- Loss of hair on the thorax
- K-wing disorder (where the rows of hooks holding the pairs of the bee's wings together become detached)

Identification

How to identify Acarapisosis

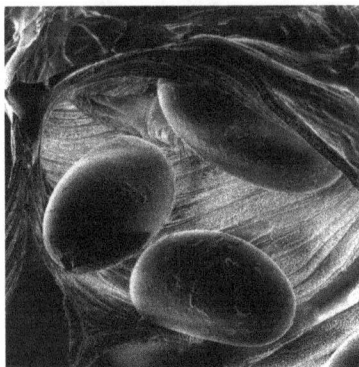

Tracheal mite (*Acarapis woodi*) eggs
Detailed view of Tracheal mite eggs within honey bee tracheal tube
Courtesy the Animal and Plant Health Agency © Crown Copyright

There are several ways to identify Acarapisosis in honeybees:

- Microscopic examination: The most definitive method to detect Acarapis woodi is to examine honeybee samples under a microscope.
- Laboratory testing: Laboratory testing can also be used to detect the presence of Acarapis woodi and other honeybee pests and diseases.
- Clinical signs: Experienced beekeepers may be able to recognise signs of Acarapisosis, such as deformed wings and stunted growth

Impact on colonies

Acarapisosis can have a significant impact on honeybee colonies. Heavy infestation can lead to the collapse of a colony.

Mitigation

There are a number of things that beekeepers can do to mitigate the impact of acarapisosis:

- Keep the hive strong and healthy.
- Inspect the hive regularly for signs of infestation.
- Treat infested hives with an approved miticide.
- Use clean and sanitised hive equipment.
- Avoid introducing new bees into the hive without first inspecting them for mites.
- Avoid using hive equipment from other hives.
- Keep the hive clean and free of debris.
- Provide the bees with a balanced diet.
- Monitor the hive for signs of stress, such as increased swarming or reduced honey production.

If you suspect that your hive is infested with acarapisosis, it is important to contact a qualified beekeeper for advice.

Additional advice on mitigation:

Use only natural or organic miticides to avoid harming the bees or the environment.
Treat infested hives in the early spring or late fall, when the mites are less active.
Repeat the treatment every 7-10 days for 3-4 treatments.
Burn or otherwise dispose of infested hive equipment.

2. ACUTE BEE PARALYSIS VIRUS (ABPV)

Acute bee paralysis virus (ABPV) is a single-stranded RNA virus that belongs to the family Dicistroviridae. It is classified as follows:

- Kingdom: Viruses
- Phylum: Riboviria
- Class: Picornaviricetes
- Order: Picornavirales
- Family: Dicistroviridae
- Genus: Dicistrovirus
- Species: Apis mellifera acute bee paralysis virus

ABPV is closely related to other honeybee viruses, such as Israeli bee virus (IBV) and Kashmir bee virus (KBV).

Biology

ABPV can infect all stages of honeybees, including eggs, larvae, pupae and adult bees. The virus is transmitted to bees through the consumption of contaminated food, such as pollen, nectar and honey. ABPV can also be transmitted to bees through contact with infected bees or their faeces.

Once inside a bee, ABPV replicates rapidly and can damage a variety of tissues, including the nervous system, digestive system and respiratory system. Infected bees may develop a variety of symptoms, including paralysis, tremors and disorientation.

Transmission

ABPV is transmitted to bees through the consumption of contaminated food, such as pollen, nectar and honey. ABPV can also be transmitted to bees through contact with infected bees or their faeces.

ABPV is a highly contagious virus and can spread quickly through a honeybee colony. The virus can also be spread from hive to hive by bees that are foraging for food.

Symptoms

The symptoms of ABPV can vary depending on the severity of the infection. Some bees may be asymptomatic, while others may develop severe symptoms and die. Common symptoms of ABPV include:

- Paralysis
- Tremors
- Disorientation
- Difficulty flying
- Reduced foraging activity
- Increased mortality

How to identify ABPV

ABPV can be identified through laboratory testing of honeybee samples. The most common methods of detection include:

- ▸ Ensyme-linked immunosorbent assay (ELISA): This is a rapid and sensitive test that can detect the presence of ABPV antibodies in honeybee samples.
- ▸ Reverse transcription-polymerase chain reaction (RT-PCR): This test can detect the presence of ABPV RNA in honeybee samples.
- ▸ Microscopic examination: In some cases, ABPV can be identified by examining honeybee tissues under a microscope.

Impact on individual bees and colonies

ABPV can have a significant impact on individual bees and colonies. Infected bees may be unable to fly, forage for food, or care for their young. In severe cases, ABPV can kill bees.

ABPV can also weaken colonies and make them more susceptible to other diseases and pests. In severe cases, ABPV can kill colonies.

Preventing and managing acute bee paralysis virus

There is no specific treatment for ABPV. However, there are steps that beekeepers can take to prevent and manage the disease:

- ▸ Maintain strong colonies: Strong colonies are more resistant to diseases and pests.
- ▸ Provide high-quality food: Beekeepers should provide their colonies with high-quality food sources, such as fresh pollen and nectar.
- ▸ Avoid stress: Stress can weaken colonies and make them more susceptible to disease. Beekeepers should avoid stressing their colonies by handling them gently and avoiding unnecessary disturbances.
- ▸ Monitor colonies for signs of disease: Beekeepers should regularly inspect their colonies for signs of disease, such as dead bees, deformed bees and diseased brood.
- ▸ Remove infected brood: Beekeepers should remove infected brood from the hive. This will help to prevent the spread of the virus.
- ▸ If you suspect that your honeybees have ABPV, you should contact your local beekeeping association for assistance.

Additional information

ABPV is a global problem and has been found in honeybees on every continent except Antarctica. The virus is particularly common in tropical and subtropical regions.

ABPV is a major concern for beekeepers because it can weaken colonies and make them more susceptible to other diseases and pests. In severe cases, ABPV can kill colonies.

Researchers are working to develop new methods for preventing and controlling ABPV. One promising approach is the development of vaccines for honeybees.

3. ACUTE BEE PARALYSIS VIRUS-LIKE VIRUS

Acute bee paralysis virus-like virus is part of the dicistroviridae family which is classified as follows

Classification

- ▸ Kingdom: Eukaryota
- ▸ Phylum: Euglenosoa
- ▸ Class: Dicistrovirida
- ▸ Order: Picornavirales
- ▸ Family: Dicistroviridae
- ▸ Genus: Dicistrovirus
- ▸ Species: Apis mellifera acute bee paralysis virus-like virus

Biology

ABPV-like viruses are classified in the genus Dicistrovirus, which also includes the related viruses Kashmir bee virus (KBV) and Egyptian acute paralysis virus (EAPV). The viruses are typically 90 nm in diameter and have a capsid composed of 180 capsomeres. ABPV-like viruses have a relatively simple genome, with only about 6.5 kb of RNA.

The life cycle of ABPV-like viruses is relatively simple and involves three main phases:

Replication in Host Cells: ABPV-like viruses enter honeybee cells and hijack the cell's machinery to replicate their own RNA genome.

Virus Assembly: During this phase, the replication machinery produces new virus particles, which assemble into mature virions.

Release: Mature virions are released from infected honeybee cells, which can then infect other bees or be spread through the hive.

Transmission

ABPV-like viruses primarily spread through two main routes:

Vertical Transmission: ABPV-like viruses can be transmitted from queen to larvae during embryonic development or during the feeding of royal jelly. This mode of transmission ensures that newborn bees are already infected from the start.

Horizontal Transmission: ABPV-like viruses can also spread horizontally within a colony through direct contact between infected and uninfected bees. This can occur through physical contact, the sharing of food sources, or through the sharing of hive equipment. Varroa mites, parasitic insects that infest honeybee colonies, can also act as vectors of ABPV-like viruses, transmitting the virus to healthy bees during their feeding process.

Symptoms of ABPV-like Virus Infection

The symptoms of ABPV-like virus infection in honeybees can vary depending on the strain of the virus, the age of the bee, and the overall health of the colony. However, some common signs include:

> ▶ Reduced Brood Viability: ABPV-like viruses can infect honeybee larvae, causing them to die prematurely or emerge with developmental abnormalities. This can lead to a decline in brood viability and a decrease in the overall colony population.
> ▶ Impaired Brood Development: The viruses can also disrupt the normal development of honeybee larvae, leading to delayed pupation and emergence time. This can further strain the colony's resources and reduce its productivity.
> ▶ Reduced Adult Bee Performance: Infected adult bees may exhibit behavioural abnormalities, such as lethargy, reduced foraging activity, and impaired learning and memory. This can significantly impact the colony's ability to gather nectar and pollen, essential for its survival.
> ▶ Increased Susceptibility to Other Pathogens: ABPV-like virus infection can weaken the immune system of honeybees, making them more susceptible to secondary infections caused by other pathogens like Varroa destructor and Nosema ceranae.

Identifying ABPV-like Virus

Diagnosis of ABPV-like virus infection in honeybees can be challenging due to the lack of specific clinical tests for individual bees. However, observing characteristic symptoms and analysing hive samples can provide clues about the presence of the virus:

> ▶ Monitoring Colonies: Monitoring colonies for signs of reduced brood viability, impaired brood development, and poor adult bee performance can raise concerns about the potential for ABPV-like virus infection.
> ▶ Hive Samples: Laboratory analysis of hive samples, such as dead bees, pollen, and brood, can detect the presence of virus particles or genetic material using molecular techniques like polymerase chain reaction (PCR).

Impact on Colonies

ABPV-like viruses can have a significant negative impact on honeybee colonies, contributing to colony decline and compromising their overall health and productivity. The viruses can disrupt brood development, weaken adult bees, and increase their susceptibility to other pathogens. These factors can lead to a reduction in colony size, reduced honey production, and an increased risk of colony collapse disorder (CCD).

Mitigation Strategies

While there is no specific treatment or cure for ABPV-like virus infection, several preventive and management strategies can be implemented to reduce the risk of outbreaks and minimise their impact:

- ▶ Strong Colony Management: Maintaining healthy, thriving colonies with adequate nutrition, ventilation, and hygienic practices can enhance the overall resistance of bees to infections.
- ▶ Monitoring and Control of Varroa Mites: Varroa mites are major vectors of ABPV-like viruses, so effective Varroa mite control is crucial for overall colony health. This can be achieved through various methods, including chemical treatments, apiary sanitation practices, and biological control.
- ▶ Protecting Hives from Pesticides: Pesticide exposure can disrupt the gut microbiome of honeybees and reduce their resistance to diseases. Avoiding or minimising pesticide use around hives can help protect bee health.
- ▶ Providing Adequate Space: Overcrowding can stress honeybees and make them more susceptible to infections. Providing ample space for colonies to grow and thrive can help reduce the risk of outbreaks.

Monitoring and Control of Varroa Mites: Varroa mites are major vectors of ABPV-like viruses, so effective Varroa mite control is crucial for overall colony health. This can be achieved through various methods, including chemical treatments, apiary sanitation practices, and biological control.

Protecting Hives from Pesticides: Pesticide exposure can disrupt the gut microbiome of honeybees and reduce their resistance to diseases. Avoiding or minimising pesticide use around hives can help protect bee health.

Providing Adequate Space: Overcrowding can stress honeybees and make them more susceptible to infections. Providing ample space for colonies to grow and thrive can help reduce the risk of outbreaks.

4. ADDLED BROOD

Addled brood is a condition in honeybee larvae or pupae that causes them to die and decompose before they can emerge as adults. It is not a specific disease, but rather a symptom of a variety of problems, including:

- ▶ Varroa mites
- ▶ American foulbrood
- ▶ European foulbrood
- ▶ Chilled brood
- ▶ Overcrowding
- ▶ Poor nutrition
- ▶ Pesticide poisoning

Biology and transmission

The biology and transmission of addled brood varies depending on the underlying cause. However, in general, addled brood is caused by a disruption in the normal development of the larvae or pupae. This disruption can be caused by a variety of factors, including:

- ▶ Varroa mites: Varroa mites are parasitic mites that attach themselves to honeybees and feed on their blood. Heavy varroa mite infestation can lead to addled brood by weakening the bees and making them more susceptible to disease.
- ▶ American foulbrood: American foulbrood is a bacterial disease that affects honeybee larvae. The bacteria produce spores that can contaminate the hive and be ingested by the larvae. The spores germinate in the larvae's gut and produce toxins that kill the larvae.
- ▶ European foulbrood: European foulbrood is a bacterial disease that also affects honeybee larvae. The bacteria produce toxins that kill the larvae.
- ▶ Chilled brood: Chilled brood is a condition that occurs when honeybee larvae or pupae are exposed to cold temperatures. The cold temperatures slow down or stop the development of the larvae or pupae, which can lead to their death.
- ▶ Overcrowding: Overcrowding can lead to addled brood by increasing the stress levels on the bees and making them more susceptible to disease.
- ▶ Poor nutrition: Poor nutrition can lead to addled brood by weakening the bees and making them more susceptible to disease.
- ▶ Pesticide poisoning: Pesticide poisoning can lead to addled brood by killing the bees or weakening them and making them more susceptible to disease.

Symptoms

The symptoms of addled brood include:

▶ Discoloured larvae or pupae (grey, brown, or black)
▶ Larvae or pupae that are swollen or misshapen
▶ Larvae or pupae that are lying on their backs or sides
▶ Larvae or pupae that have a foul odour

How to identify addled brood

Addled brood can be identified by inspecting the brood comb for the presence of dead larvae in capped cells. The larvae will have the characteristic sunken, discoloured appearance.

If you suspect that you have addled brood in your colony, it is important to have it inspected by a bee expert to determine the cause of the problem.

Impact on colonies

Addled brood can have a significant impact on honeybee colonies. Heavily addled brood can lead to a decrease in the number of adult bees in the colony, which can reduce honey production and make the colony more susceptible to other problems.

Mitigation

The best way to mitigate the impact of addled brood is to identify and address the underlying cause. For example, if varroa mites are the cause of the addled brood, then the beekeeper should treat the hive with an approved miticide. If American foulbrood is the cause of the addled brood, then the beekeeper should burn the hive and all the equipment associated with the hive.

In addition to addressing the underlying cause, beekeepers can also help to reduce the risk of addled brood by:

▶ Keeping the hive strong and healthy.
▶ Inspecting the hive regularly for signs of disease.
▶ Providing the bees with a balanced diet.
▶ Avoid overcrowding the hive.
▶ Using only approved pesticides.

If you suspect that your hive has addled brood, it is important to contact a qualified beekeeper for assistance.

5. ADENOVIRUS

In honeybees, adenovirus are relatively common viruses and are prevalent in some 70% of colonies in some populations.

- ▸ Kingdom: Viruses
- ▸ Realm: Riboviria
- ▸ Phylum: Negarnaviricota
- ▸ Class: Wichivirus
- ▸ Order: Yersiniales
- ▸ Family: Adenoviridae
- ▸ Genus: Mastadenovirus

Biology

Adenoviruses are non-enveloped DNA viruses that replicate in the nucleus of infected cells. They have a wide host range, including humans, animals and birds. Honeybees are susceptible to several types of adenoviruses, including Adenovirus Apis mellifera (AaMV).

AaMV replicates in the nucleus of infected honeybee cells, causing damage to the cells and eventually killing them. The virus is transmitted horizontally from bee to bee through contact with contaminated food, water, or hive equipment. It can also be transmitted by Varroa mites, which are parasitic mites that feed on honeybees.

Transmission

AaMV is primarily transmitted through contact with contaminated food or water. Infected bees can shed the virus through their faeces or saliva, which can contaminate food and water sources for other bees. AaMV can also be transmitted through contact with contaminated hive equipment, such as frames, combs and honey pots. Varroa mites can also transmit AaMV to honeybees by feeding on infected bees and then transmitting the virus to healthy bees.

Symptoms

The symptoms of AaMV infection in honeybees can vary depending on the strain of virus involved and the severity of the infection. Some of the most common symptoms include:

- ▸ Reduced honey production
- ▸ Increased mortality
- ▸ Deformed wings
- ▸ Paralysis
- ▸ Hair loss
- ▸ Darkened abdomens
- ▸ Stunted cell growth
- ▸ Abnormal bee behaviour

How to identify AaMV

AaMV can be identified by laboratory testing of honeybee samples. The most common methods of detection include:

Reverse transcription-polymerase chain reaction (RT-PCR): This test can detect the presence of AaMV DNA in honeybee samples.

Ensyme-linked immunosorbent assay (ELISA): This test can detect the presence of AaMV antibodies in honeybee samples.

Impact on colonies

AaMV infection can have a significant impact on honeybee colonies. Heavy infection can lead to:
- Reduced honey production
- Increased mortality
- Colony collapse

Advice on mitigation

There is no cure for AaMV infection in honeybees. However, there are a number of things that beekeepers can do to mitigate the impact of the virus:
- Maintain strong colony health: Strong colonies are better able to withstand the effects of AaMV infection.
- Provide adequate nutrition: Ensure that the colony has access to a plentiful supply of pollen and other essential nutrients.
- Control hive temperature: Avoid extreme temperatures in the hive by providing adequate ventilation and insulation.
- Avoid pesticide exposure: Use pesticides only as a last resort and follow label directions carefully.
- Monitor colony health: Regularly inspect the colony for signs of AaMV infection, such as reduced honey production, increased mortality, deformed wings, paralysis, hair loss, darkened abdomens, stunted cell growth and abnormal bee behaviour.
- Isolate infected colonies: If you find AaMV infection in a colony, isolate it from other colonies to prevent the spread of the virus.
- Destroy infected hives: If a colony is heavily infected with AaMV, it is best to destroy the hive and all its contents.
- Purchase bees from reputable beekeepers: Purchase bees from reputable beekeepers who have taken steps to control AaMV in their colonies.

6. AETHINA TUMIDA

See small hive beetle

7. AMERICAN FOUL BROOD

American foulbrood (AFB) is a bacterial disease of honeybee larvae caused by the spore-forming bacterium Paenibacillus larvae. It is classified as follows:

- ▸ Kingdom: Bacteria
- ▸ Phylum: Firmicutes
- ▸ Class: Bacilli
- ▸ Order: Bacillales
- ▸ Family: Paenibacillaceae
- ▸ Genus: Paenibacillus
- ▸ Species: P. larvae

Biology

American Foulbrood (AFB)
AFB infected larvae forming a
drawn out, viscous string during a
rope test

Courtesy the Animal and Plant
Health Agency © Crown Copyright

Paenibacillus larvae is a rod-shaped bacterium that measures approximately 0.5-1.0 μm in diameter and 1.0-6.0 μm in length. It is a highly resistant bacterium, forming spores that can survive for decades in the environment.

The life cycle of Paenibacillus larvae is as follows:

Vegetative growth: The spores germinate and grow into vegetative bacteria within the honeybee larvae.

Multiplication: The vegetative bacteria multiply rapidly and invade the larval tissues.

Sporulation: As the larvae die, the vegetative bacteria convert back into spores.

Release of spores: The spores are released into the environment when the larval cappings are broken or when the dead larvae are removed from the hive.

Transmission

American foulbrood is primarily transmitted to honeybees through contact with contaminated honeycombs, honey and hive equipment. Spores can also be transmitted through contaminated pollen, water and soil.

Symptoms

The symptoms of American foulbrood in honeybees can vary depending on the severity of the infection. Some of the most common symptoms include:

- ▶ Irregular and patchy brood pattern: The brood pattern may become irregular and patchy due to the death of infected larvae.
- ▶ Sunken, discoloured and greasy cappings: The cappings of cells containing dead larvae may become sunken, discoloured and greasy.
- ▶ Rope test: The contents of infected cells may be ropey and sticky when pulled out with a toothpick.
- ▶ Dead larvae: Dead larvae may appear sunken, discoloured and have a foul odour.
- ▶ Scale formation: The remains of dead larvae may dry and form scales that adhere to the cell walls.
- ▶ Reduced honey production: Infected colonies may produce less honey due to the weakened condition of the bees.
- ▶ Increased mortality: Heavily infested colonies may experience increased mortality rates among bees.
- ▶ Colony collapse: In severe cases, American foulbrood can lead to the collapse of the entire colony.

How to identify American foulbrood

American foulbrood can be identified by examining the brood comb for the presence of characteristic symptoms, such as sunken, discoloured cappings, ropey larvae and dead larvae with scales. Laboratory testing can also be used to confirm the diagnosis along with lateral flow devices.

Impact on colonies

AFB can have a devastating impact on honeybee colonies. Heavy infestation can lead to the collapse of a colony within a few weeks. AFB can also weaken the colony and make it more susceptible to other problems.

American Foulbrood (AFB)
Comb severely infected with
American Foulbrood (AFB)

Courtesy the Animal and Plant
Health Agency © Crown Copyright

Mitigation

There is no cure for AFB. The only way to control the disease is to destroy the infected hive and all the equipment associated with the hive.

- ▶ Beekeepers can help to prevent AFB by:
- ▶ Inspecting the hive regularly for signs of disease.
- ▶ Keeping the hive strong and healthy.

- ▶ Using clean and sanitised hive equipment.
- ▶ Avoiding overcrowding the hive.
- ▶ Avoiding introducing new bees into the hive without first inspecting them for disease.
- ▶ Purchasing bees from reputable beekeepers.

If you suspect that your hive has AFB, it is important to contact a qualified beekeeper for assistance.

Additional advice on mitigation:

- ▶ Burn the hive and all the equipment associated with the hive.
- ▶ Clean and disinfect the apiary.
- ▶ Do not reuse hive equipment from an infected hive.
- ▶ Notify other beekeepers in the area.
- ▶ Report the AFB outbreak to the authorities.

Additional research

Researchers are working to develop new ways to control AFB. One promising approach is to use bacteriophages, which are viruses that infect bacteria. Bacteriophages can be used to kill the AFB bacterium without harming the bees.

Another promising approach is to develop genetically resistant bees. Bees can be bred to have genes that make them resistant to the AFB bacterium.

By continuing to research AFB, scientists can develop new and effective ways to control the disease and protect honeybee colonies.

8. AMOEBOSIS

Amoebosis in honeybees is caused by the parasitic protosoan Malpighamoeba mellificae. It is classified as follows:

- Kingdom: Protosoa
- Phylum: Amoebosoa
- Class: Lobosea
- Order: Malpighamonadida
- Family: Malpighamoebidae
- Genus: Malpighamoeba
- Species: M. mellificae

Biology and transmission

Malpighamoeba mellificae infects the Malpighian tubules of honeybees. The Malpighian tubules are excretory organs that are responsible for removing waste products from the bee's body.

The amoeba is transmitted from bee to bee through contact with infected bees or their faeces. The amoeba can also be transmitted through contaminated hive equipment.

Once inside the bee, the amoeba multiplies and damages the Malpighian tubules. This can lead to a variety of problems, including:

- Reduced honey production
- Increased swarming
- Decreased lifespan
- Death

Symptoms

The symptoms of amoebosis in honeybees are not always obvious. However, some common symptoms include:

- Diarrhoea
- Dysentery
- Death

How to identify Amoebosis

Amoebosis can be identified by microscopic examination of the Malpighian tubules of infected bees. Laboratory testing can also be used to detect the presence of Malpighamoeba mellificae.

Impact on colonies

Amoebosis can have a significant impact on honeybee colonies. Heavy infestation can lead to the collapse of a colony. Amoebosis can also weaken the colony and make it more susceptible to other problems.

Mitigation

There is no cure for amoebosis in honeybees. However, there are a number of things that beekeepers can do to mitigate the impact of the disease:

▶ Keep the hive strong and healthy.
▶ Inspect the hive regularly for signs of disease.
▶ Avoid overcrowding the hive.
▶ Use clean and sanitised hive equipment.
▶ Avoid introducing new bees into the hive without first inspecting them for disease.
▶ Purchasing bees from reputable beekeepers.

If you suspect that your hive has amoebosis, it is important to contact a qualified beekeeper for assistance.

Additional advice on mitigation:

Provide the bees with a balanced diet.

Monitor the hive for signs of stress, such as increased swarming or reduced honey production.

Treat the hive with an approved miticide to control varroa mites, which can weaken the bees and make them more susceptible to amoebosis.

9. APHID LETHAL PARALYIS VIRUS

Aphid lethal paralysis virus is a single-stranded, negative-sense RNA virus classified as follows:

Classification:

- ▶ Kingdom: Riboviria
- ▶ Phylum: Negarnaviricota
- ▶ Class: Monjiviricetes
- ▶ Order: Dicistroviridae
- ▶ Family: Dicitroviridae
- ▶ Genus: Aphidlethalparalysisvirus
- ▶ Species: Aphid lethal paralysis virus

Biology:

ALPV is a single-stranded, negative-sense RNA virus belonging to the Dicistroviridae family. It primarily replicates within aphid species, causing paralysis and death. However, certain strains of ALPV (mainly ALPV-Am) can also infect honeybees, although their effects are not fully understood.

Transmission:

Direct contact: Infected aphids can transmit ALPV to honeybees through physical contact during feeding or interactions within the hive.

Contaminated pollen: Honeybees can ingest ALPV-contaminated pollen, potentially introducing the virus into the colony.

Symptoms:

Deformed wings: ALPV infection can lead to deformed wings in adult honeybees, similar to Deformed Wing Virus (DWV).

Shortened legs: Like other bee viruses, ALPV may cause stunted leg development.

Other deformities: Although less common, ALPV can cause other deformities in bees, such as misshapen abdomens, missing limbs, and eye abnormalities.

Reduced brood production: Infected colonies may experience decreased brood production due to developmental issues or viral effects on queen bee health.

Increased mortality: Heavy ALPV infection can contribute to a rise in adult bee mortality.

Identification:

Clinical symptoms: Observing deformed bees and reduced brood production can raise suspicion of ALPV infection, but definitive diagnosis requires laboratory testing, such as RT-PCR.

Aphid presence: High aphid populations on hives or frames suggest a potential source of ALPV transmission.

Impact on Colonies:

While ALPV's impact on honeybees is still being researched, it can contribute to colony weakening through:

▶ Decreased foraging abilities: Deformed wings and paralysis hinder bees' ability to collect pollen and nectar, impacting colony resources.

▶ Colony instability: Reduced brood production and higher mortality can destabilise colony population and long-term survival.

▶ Potential co-infections: ALPV may interact with other honeybee viruses like DWV, potentially aggravating their effects.

Mitigation:

Integrated Pest Management (IPM): Controlling aphid populations around hives can decrease direct ALPV transmission to bees.

Hive hygiene: Maintaining clean and well-ventilated hives can discourage aphid infestations and limit viral spread.

Monitoring: Routine hive inspections and monitoring for deformed bees and reduced brood production can help detect potential ALPV infection early.

Research advancements: Ongoing research on ALPV, its transmission, and potential effects on honeybees will further inform mitigation strategies.

10. APISMELLIFERA SOLINVIVIRUS TYPE 1

Apis mellifera solinvivirus (AmSV1) is a single-stranded RNA virus that belongs to the family Solinviviridae. It is classified as follows:

Classification

- Kingdom: Viruses
- Phylum: Riboviria
- Class: Picornaviricetes
- Order: Picornavirales
- Family: Solinviviridae
- Genus: Solinvivirus
- Species: Apis mellifera solinvivirus

Biology and transmission

AmSV1 is a relatively new virus that was first discovered in honeybees in the United States in 2013. The virus is transmitted from bee to bee through contact with infected bees or their faeces. AmSV1 can also be transmitted through contaminated hive equipment.

Once inside the bee, AmSV1 replicates in the bee's gut and nervous system. The virus can cause paralysis and death in bees.

Symptoms

The symptoms of AmSV1 infection in honeybees are not always obvious. However, some common symptoms include:

- Paralysis
- Trembling
- Death

Impact on colonies

AmSV1 infection can have a significant impact on honeybee colonies. Heavy infestation can lead to the collapse of a colony. AmSV1 infection can also weaken the colony and make it more susceptible to other problems.

How to identify the virus

AmSV1 infection can be identified by testing honeybee samples for the presence of the virus's RNA. This can be done using a variety of methods, such as real-time PCR (qPCR) or reverse transcription PCR (RT-PCR).

Mitigation

There is no cure for AmSV1 infection in honeybees. However, there are a number of things that beekeepers can do to mitigate the impact of the disease:

- Keep the hive strong and healthy.

- ▸ Inspect the hive regularly for signs of disease.
- ▸ Avoid overcrowding the hive.
- ▸ Use clean and sanitised hive equipment.
- ▸ Avoid introducing new bees into the hive without first inspecting them for disease.
- ▸ Purchasing bees from reputable beekeepers.

If you suspect that your hive has AmSV1 infection, it is important to contact a qualified beekeeper or veterinarian for assistance.

Additional advice on mitigation:

Provide the bees with a balanced diet.

Monitor the hive for signs of stress, such as increased swarming or reduced honey production.

Treat the hive with an approved miticide to control varroa mites, which can weaken the bees and make them more susceptible to AmSV1 infection.

Research on AmSV1 is ongoing and scientists are working to develop new ways to control the virus and protect honeybee colonies.

11. ASCOPHAERA APIS (CHALK BROOD)

Ascophaera apis is a fungus that causes chalkbrood disease in honeybees. It is classified as follows:

Classification

▶ Kingdom: Fungi
▶ Phylum: Ascomycota
▶ Class: Leotiomycetes
▶ Order: Helotiales
▶ Family: Ascosphaeraceae
▶ Genus: Ascophaera
▶ Species: Ascophaera apis

Biology

European Foulbrood (EFB) melted down larvae
European Foulbrood infected larvae at various stages of decay

Courtesy the Animal and Plant Health Agency © Crown Copyright

Ascophaera apis is a microscopic fungus that produces spores that are resistant to drying and can survive in the environment for many years. The spores are ingested by honeybee larvae when they are fed contaminated food by nurse bees. The spores germinate in the larvae's gut and produce mycelium, which grows and kills the larvae.

The mummified remains of the larvae are then covered in a white powder, which is the fungus's fruiting bodies. These fruiting bodies produce more spores, which continue the cycle of infection.

Transmission

Ascophaera apis is transmitted from bee to bee through contact with contaminated food, hive equipment, or the bodies of dead and infected larvae. The fungus can also be transmitted by adult bees that carry spores on their bodies.

Symptoms

The symptoms of chalkbrood disease in honeybees include:

▶ Mummified larvae that are covered in a white powder
▶ Discoloured larvae (light brown to dark brown or black)
▶ Larvae that are dead and ropy
▶ A foul odour from the hive
▶ Decreased honey production
▶ Increased swarming

▸ Colony collapse

How to identify the fungus

Ascophaera apis can be identified by examining the mummified remains of larvae. The fungus's fruiting bodies are visible as a white powder on the surface of the mummies. The fungus can also be identified by microscopic examination of the mummies.

Impact on colonies

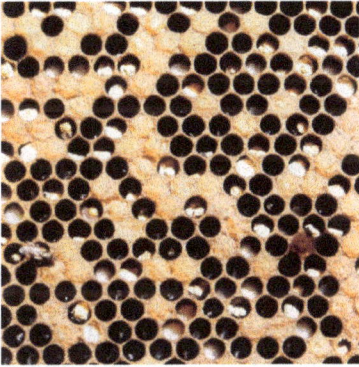

European Foulbrood (EFB)
Advance stage of EFB infection in comb with "melted", discoloured, slumped larvae.
courtesy Animal and Plant Health Agency © Crown copyright

Chalkbrood disease can have a significant impact on honeybee colonies. Heavy infestation can lead to the collapse of a colony. Chalkbrood disease can also weaken the colony and make it more susceptible to other problems, such as varroa mite infestation and American foulbrood.

Mitigation

There is no cure for chalkbrood disease. However, there are a number of things that beekeepers can do to mitigate the impact of the disease:

▸ Keep the hive strong and healthy.
▸ Inspect the hive regularly for signs of disease.
▸ Avoid overcrowding the hive.
▸ Use clean and sanitised hive equipment.
▸ Avoid introducing new bees into the hive without first inspecting them for disease.
▸ Purchasing bees from reputable beekeepers.
▸ Treat the hive with an approved miticide to control varroa mites, which can weaken the bees and make them more susceptible to chalkbrood disease.
▸ Provide the bees with a balanced diet.
▸ Monitor the hive for signs of stress, such as increased swarming or reduced honey production.

Additional advice on mitigation:

▸ Remove any dead or diseased larvae from the hive as soon as possible.
▸ Burn or otherwise dispose of contaminated hive equipment.
▸ Clean and disinfect the apiary.
▸ Get support from your local beekeepers.

Researchers are working to develop new ways to control chalkbrood disease. One promising approach is to develop fungicides that are effective against Ascophaera apis

12. ASPERGILLUS FLAVUS

Aspergillus flavus is a cosmopolitan fungus that is classified as follows:

Classification

- Kingdom: Fungi
- Phylum: Ascomycota
- Class: Eurotiomycetes
- Order: Eurotiales
- Family: Trichocomaceae
- Genus: Aspergillus
- Species: Aspergillus flavus

Biology

Aspergillus flavus is a cosmopolitan fungus that is found in soil, air and food. It is a common contaminant of food and can produce toxins that can cause illness in humans and animals.

Aspergillus flavus can infect honeybees at all stages of development, from larvae to adults. The fungus can enter the hive through contaminated food, pollen, or hive equipment.

Once inside the hive, Aspergillus flavus can infect the bees' gut and nervous system. The fungus can also cause disease in the bees' brood.

Transmission

Aspergillus flavus is transmitted from bee to bee through contact with contaminated food, hive equipment, or the bodies of dead and infected bees. The fungus can also be transmitted by adult bees that carry spores on their bodies.

Symptoms

The symptoms of Aspergillus flavus infection in honeybees include:
- Discoloured larvae
- Dead and ropy larvae
- A foul odour from the hive
- Decreased honey production
- Increased swarming
- Colony collapse

How to identify the fungus

Aspergillus flavus can be identified by microscopic examination of samples from the hive, such as brood, food and hive equipment. The fungus can also be identified by culturing samples on specialised media.

Impact on colonies

Aspergillus flavus infection can have a significant impact on honeybee colonies. Heavy infestation can lead to the collapse of a colony. Aspergillus flavus infection can also weaken the colony and make it more susceptible to other problems, such as varroa mite infestation and American foulbrood.

Mitigation

There is no cure for Aspergillus flavus infection in honeybees. However, there are a number of things that beekeepers can do to mitigate the impact of the disease:

- ▶ Keep the hive strong and healthy.
- ▶ Inspect the hive regularly for signs of disease.
- ▶ Avoid overcrowding the hive.
- ▶ Use clean and sanitised hive equipment.
- ▶ Avoid introducing new bees into the hive without first inspecting them for disease.
- ▶ Purchasing bees from reputable beekeepers.
- ▶ Treat the hive with an approved fungicide to control Aspergillus flavus.
- ▶ Provide the bees with a balanced diet.
- ▶ Monitor the hive for signs of stress, such as increased swarming or reduced honey production.

Additional advice on mitigation:

- ▶ Remove any dead or diseased bees from the hive as soon as possible.
- ▶ Burn or otherwise dispose of contaminated hive equipment.
- ▶ Clean and disinfect the apiary.

Researchers are working to develop new ways to control Aspergillus flavus infection in honeybees. One promising approach is to develop fungicides that are more effective against the fungus. Another promising approach is to develop bee strains that are resistant to Aspergillus flavus infection.

13. ASPERGILLUS FUMIGATUS

This is a fungal disease that affects honeybees and is classified as follows:

Classification

- ▶ Kingdom: Fungi
- ▶ Phylum: Ascomycota
- ▶ Class: Eurotiomycetes
- ▶ Order: Eurotiales
- ▶ Family: Trichocomaceae
- ▶ Genus: Aspergillus
- ▶ Species: Aspergillus fumigatus

Biology

Aspergillus fumigatus is a filamentous fungus that can grow in various environmental conditions, including soil, decaying plant matter, and honeybee colonies. The fungus produces spores that can be spread through the air, water, or direct contact with contaminated surfaces

Transmission

Aspergillus fumigatus spores are ubiquitous in the environment and can enter honeybee colonies through various routes, including:

Airborne Transmission: Spores can be carried into the hive by wind or bees returning from foraging.

Pollen and Nectar: Spores can contaminate pollen and nectar, which can then be brought into the hive by foraging bees.

Varroa Mites: Varroa mites can pick up spores from contaminated pollen or nectar and then transfer them to honeybees during their parasitic feeding.

Symptoms of Mucormycosis

- ▶ Honeybees infected with Aspergillus fumigatus may exhibit the following symptoms:
- ▶ White, cottony fungal growth on the bodies of bees.
- ▶ Ragged wings, making it difficult for bees to fly.
- ▶ Loss of coordination and balance.
- ▶ Clusters of dead bees in or near the hive.
- ▶ Reduced colony size.

How to identify Aspergillus fumigatus

Identifying mucormycosis in honeybees can be challenging, as the symptoms are often similar to those of other diseases. However, the presence of white, cottony fungal growth on bees can be a strong indicator of the disease. Laboratory testing of hive samples, such as dead bees or pollen, can confirm the presence of Aspergillus fumigatus spores or fungal hyphae.

Impact of Aspergillus fumigatus on colonies

Mucormycosis can have a significant negative impact on honeybee colonies, leading to colony decline and reduced productivity. The disease can cause the death of adult bees, reduce brood viability, and impair the ability of bees to forage and collect nectar and pollen. These factors can contribute to a decline in the overall health and productivity of honeybee colonies.

Advice on Mitigation

There are a number of measures that beekeepers can take to reduce the risk of aspergillus fumigatus outbreaks and minimise their impact:

▶ Maintaining Hygienic Hive Conditions: Good hive sanitation practices can help to reduce the number of fungal spores present in the hive. This includes regular cleaning of hive equipment, removing dead bees and debris, and ensuring adequate ventilation.

▶ Controlling Varroa Mites: Varroa mites can act as vectors for aspergillus fumigatus, so controlling their populations is important for reducing the risk of the disease. This can be achieved through the use of various miticides or integrated pest management (IPM) strategies.

▶ Monitoring Pollen and Nectar Sources: Beekeepers should be aware of pollen and nectar sources that may be contaminated with Aspergillus fumigatus spores. This may involve avoiding certain areas or times of the year when spore levels are high.

▶ Protecting Colonies from Stress: Stressful conditions, such as poor nutrition, overcrowding, or pesticide exposure, can weaken honeybee colonies and make them more susceptible to disease. Beekeepers should strive to provide their colonies with adequate resources and avoid exposing them to unnecessary stressors.

▶ Using Fungal-Resistant Strains of Honeybees: Research is ongoing into developing honeybee strains with increased resistance to Aspergillus fumigatus. These strains could be used to help protect honeybee colonies from the disease.

14. ASPERGILLUS NIGER

Aspergillus niger is a cosmopolitan fungus that is classified as follows:

Classification

- ‣ Kingdom: Fungi
- ‣ Phylum: Ascomycota
- ‣ Class: Eurotiomycetes
- ‣ Order: Eurotiales
- ‣ Family: Trichocomaceae
- ‣ Genus: Aspergillus
- ‣ Species: Aspergillus niger

Biology

Aspergillus niger is a cosmopolitan fungus that is found in soil, air and food. It is a common contaminant of food and can produce toxins that can cause illness in humans and animals.

Aspergillus niger can infect honeybees at all stages of development, from larvae to adults. The fungus can enter the hive through contaminated food, pollen, or hive equipment.

Once inside the hive, Aspergillus niger can infect the bees' gut and nervous system. The fungus can also cause disease in the bees' brood.

Transmission

Aspergillus niger is transmitted from bee to bee through contact with contaminated food, hive equipment, or the bodies of dead and infected bees. The fungus can also be transmitted by adult bees that carry spores on their bodies.

Symptoms

The symptoms of Aspergillus niger infection in honeybees include:

- ‣ Discoloured larvae
- ‣ Dead and ropy larvae
- ‣ A foul odour from the hive
- ‣ Decreased honey production
- ‣ Increased swarming
- ‣ Colony collapse

How to identify the fungus

Aspergillus niger can be identified by microscopic examination of samples from the hive, such as brood, food and hive equipment. The fungus can also be identified by culturing samples on specialised media.

Impact on colonies

Aspergillus niger infection can have a significant impact on honeybee colonies. Heavy infestation can lead to the collapse of a colony. Aspergillus niger infection can also weaken the colony and make it more susceptible to other problems, such as varroa mite infestation and American foulbrood.

Mitigation

There is no cure for Aspergillus niger infection in honeybees. However, there are a number of things that beekeepers can do to mitigate the impact of the disease:

▶ Keep the hive strong and healthy.
▶ Inspect the hive regularly for signs of disease.
▶ Avoid overcrowding the hive.
▶ Use clean and sanitised hive equipment.
▶ Avoid introducing new bees into the hive without first inspecting them for disease.
▶ Purchasing bees from reputable beekeepers.
▶ Treat the hive with an approved fungicide to control Aspergillus niger.
▶ Provide the bees with a balanced diet.
▶ Monitor the hive for signs of stress, such as increased swarming or reduced honey production.

Additional advice on mitigation:

▶ Remove any dead or diseased bees from the hive as soon as possible.
▶ Burn or otherwise dispose of contaminated hive equipment.
▶ Clean and disinfect the apiary.

Researchers are working to develop new ways to control Aspergillus niger infection in honeybees. One promising approach is to develop fungicides that are more effective against the fungus. Another promising approach is to develop bee strains that are resistant to Aspergillus niger infection

15. ASTROVIRUS

Astrovirus are non-enveloped RNA viruses which are classified as follows

Classification

- ▶ Kingdom: Viruses
- ▶ Realm: Riboviria
- ▶ Phylum: Negarnaviricota
- ▶ Class: Picornaviricetes
- ▶ Order: Picornavirales
- ▶ Family: Astroviridae
- ▶ Genus: Astrovirus

Biology

Astroviruses are non-enveloped RNA viruses that replicate in the cytoplasm of infected cells. Honeybees are susceptible to several types of astroviruses, including Sacbrood virus (SBV).

Astroviruses replicate in the cytoplasm of infected honeybee larvae, causing damage to the cells and eventually killing them. The virus is transmitted horizontally from bee to bee through contact with contaminated food, water, or hive equipment. It can also be transmitted by Varroa mites, which are parasitic mites that feed on honeybees.

Transmission

Astrovirus is primarily transmitted through contact with contaminated food or water. Infected adult bees can shed the virus through their faeces or saliva, which can contaminate food and water sources for larvae. Astrovirus can also be transmitted through contact with contaminated hive equipment, such as frames, combs and honey pots. Varroa mites can also transmit Astrovirus to honeybee larvae by feeding on infected larvae and then transmitting the virus to healthy larvae.

Symptoms

The symptoms of Astrovirus infection in honeybee larvae can vary depending on the strain of virus involved and the severity of the infection. Some of the most common symptoms include:

- ▶ Reduced brood production
- ▶ Deformed larvae
- ▶ Dead larvae
- ▶ Sticky larvae

How to identify astrovirus

SBV can be identified by laboratory testing of honeybee samples. The most common methods of detection include:

- ▸ Reverse transcription-polymerase chain reaction (RT-PCR): This test can detect the presence of Astrovirus RNA in honeybee samples.
- ▸ Ensyme-linked immunosorbent assay (ELISA): This test can detect the presence of SBV antibodies in honeybee samples.

Impact on colonies

SBV infection can have a significant impact on honeybee colonies. Heavy infection can lead to:

- ▸ Reduced honey production
- ▸ Increased mortality
- ▸ Colony collapse

Advice on mitigation

- ▸ There is no cure for Astrovirus infection in honeybees. However, there are a number of things that beekeepers can do to mitigate the impact of the virus:
- ▸ Maintain strong colony health: Strong colonies are better able to withstand the effects of Astrovirus infection.
- ▸ Provide adequate nutrition: Ensure that the colony has access to a plentiful supply of pollen and other essential nutrients.
- ▸ Control hive temperature: Avoid extreme temperatures in the hive by providing adequate ventilation and insulation.
- ▸ Avoid pesticide exposure: Use pesticides only as a last resort and follow label directions carefully.
- ▸ Monitor colony health: Regularly inspect the colony for signs of Astrovirus infection, such as reduced brood production, deformed larvae, dead larvae and sticky larvae.
- ▸ Isolate infected colonies: If you find Astrovirus infection in a colony, isolate it from other colonies to prevent the spread of the virus.
- ▸ Destroy infected hives: If a colony is heavily infected with Astrovirus, it is best to destroy the hive and all its contents.
- ▸ Purchase bees from reputable beekeepers: Purchase bees from reputable beekeepers who have taken steps to control Astrovirus in their colonies.
- ▸ Monitor Varroa mite infestation: Control Varroa mite infestation as Varroa mites can transmit Astrovirus.

16. BACILLUS ALVEI

See paenibacillus larvae.

17. BACILLUS CEREUS

Bacillus cereus is a spore-forming bacterium that is classified as follows:

Classification

- ▶ Kingdom: Bacteria
- ▶ Phylum: Firmicutes
- ▶ Class: Bacilli
- ▶ Order: Bacillales
- ▶ Family: Bacillaceae
- ▶ Genus: Bacillus
- ▶ Species: Bacillus cereus

Biology

Bacillus cereus is a spore-forming bacterium, which means that it can produce spores that are resistant to heat, drying and other harsh conditions. Spores can survive for many years in the environment.

Bacillus cereus is found in soil, air and water. It is also a common contaminant of food. Bacillus cereus can infect honeybees at all stages of development, from larvae to adults. The bacteria can enter the hive through contaminated food, pollen, or hive equipment. Once inside the hive, Bacillus cereus can infect the bees' gut and nervous system. The bacteria can also cause disease in the bees' brood.

Transmission

Bacillus cereus is transmitted from bee to bee through contact with contaminated food, hive equipment, or the bodies of dead and infected bees. The bacteria can also be transmitted by adult bees that carry spores on their bodies.

Bacillus cereus can also be transmitted to humans through the consumption of contaminated honey.

Symptoms

The symptoms of Bacillus cereus infection in honeybees include:

- ▶ Diarrhoea
- ▶ Dysentery
- ▶ Death

How to identify the virus

Bacillus cereus can be identified by culturing samples from the hive, such as brood, food and hive equipment, on specialised media. The bacteria can also be identified using molecular methods, such as PCR.

Impact on colonies

Bacillus cereus infection can have a significant impact on honeybee colonies. Heavy infestation can lead to the collapse of a colony. Bacillus cereus infection can also weaken the colony and make it more susceptible to other problems, such as varroa mite infestation and American foulbrood.

Mitigation

There is no cure for Bacillus cereus infection in honeybees. However, there are a number of things that beekeepers can do to mitigate the impact of the disease:
- Keep the hive strong and healthy.
- Inspect the hive regularly for signs of disease.
- Avoid overcrowding the hive.
- Use clean and sanitised hive equipment.
- Avoid introducing new bees into the hive without first inspecting them for disease.
- Purchasing bees from reputable beekeepers.
- Avoid feeding the bees honey from unknown sources.
- Harvest honey from the hive as soon as possible after it is capped.
- Store honey in a cool, dark place.

Additional advice on mitigation:
- Remove any dead or diseased bees from the hive as soon as possible.
- Burn or otherwise dispose of contaminated hive equipment.
- Clean and disinfect the apiary.

Researchers are working to develop new ways to control Bacillus cereus infection in honeybees. One promising approach is to develop probiotics that can compete with Bacillus cereus in the bees' gut. Another promising approach is to develop vaccines that can protect the bees from Bacillus cereus infection.

18. BACILLUS MEGATERIUM

Bacillus megaterium is a large, rod-shaped bacterium that is gram-positive and spore-forming bacterium that is classified as follows:

Classification

▸ Kingdom: Bacteria
▸ Phylum: Firmicutes
▸ Class: Bacilli
▸ Order: Bacillales
▸ Family: Bacillaceae
▸ Genus: Bacillus
▸ Species: Bacillus megaterium

Biology

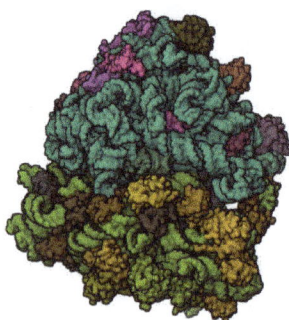

BACILLUS SPP.
COURTESY RCSB.ORG
6HA1 Structural basis for antibiotic resistance mediated by theBacillus subtilisABCF ATPase VmlR.

Crowe-McAuliffe, C., Graf, M., Huter, P., Takada, H., Abdelshahid, M., Novacek, J., Murina, V., Atkinson, G.C., Hauryliuk, V., Wilson, D.N.

(2018) Proc Natl Acad Sci U S A 115: 8978-8983

Bacillus megaterium is a large, rod-shaped bacterium that is gram-positive and spore-forming. Spores are resistant to heat, drying and other harsh conditions and can survive for many years in the environment.

Bacillus megaterium is a common inhabitant of soil and water and is also found in food and other organic materials. It is not considered to be a human pathogen, but can cause disease in some animals, including honeybees.

Transmission

Bacillus megaterium can be transmitted to honeybees through contaminated food, water, or hive equipment. It can also be transmitted by adult bees that carry spores on their bodies.

Once inside the hive, Bacillus megaterium can infect the bees' gut and nervous system. It can also cause disease in the bees' brood.

Symptoms

The symptoms of Bacillus megaterium infection in honeybees include:

▸ Diarrhoea
▸ Dysentery
▸ Death

How to identify the bacteria

Bacillus megaterium can be identified by culturing samples from the hive, such as brood, food and hive equipment, on specialised media. The bacteria can also be identified using molecular methods, such as PCR.

Impact on colonies

Bacillus megaterium infection can have a significant impact on honeybee colonies. Heavy infestation can lead to the collapse of a colony. Bacillus megaterium infection can also weaken the colony and make it more susceptible to other problems, such as varroa mite infestation and American foulbrood.

Mitigation

There is no cure for Bacillus megaterium infection in honeybees. However, there are a number of things that beekeepers can do to mitigate the impact of the disease:

▶ Keep the hive strong and healthy.
▶ Inspect the hive regularly for signs of disease.
▶ Avoid overcrowding the hive.
▶ Use clean and sanitised hive equipment.
▶ Avoid introducing new bees into the hive without first inspecting them for disease.
▶ Purchase bees from reputable beekeepers.
▶ Avoid feeding the bees honey from unknown sources.
▶ Harvest honey from the hive as soon as possible after it is capped.
▶ Store honey in a cool, dark place.

Additional advice on mitigation:

▶ Remove any dead or diseased bees from the hive as soon as possible.
▶ Burn or otherwise dispose of contaminated hive equipment.
▶ Clean and disinfect the apiary.

Researchers are working to develop new ways to control Bacillus megaterium infection in honeybees. One promising approach is to develop probiotics that can compete with Bacillus megaterium in the bees' gut. Another promising approach is to develop vaccines that can protect the bees from Bacillus megaterium infection.

19. BALD BROOD

Bald brood, a telltale sign of wax moth infestation, is not a disease itself but rather a symptom observed in honeybee colonies. It presents as patchy areas within the brood combs where sealed pupae have been uncapped, exposing their vulnerable bodies.

Classification and Biology:

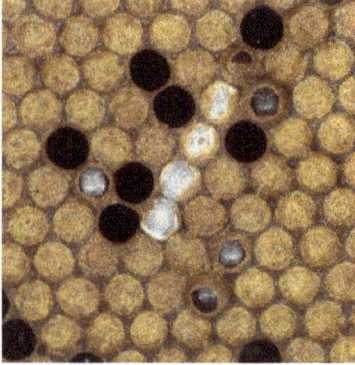

Bald Brood with Wax Moth larvae activity.
Comb showing signs of bald brood with Wax Moth larvae activity
Courtesy the Animal and Plant Health Agency © Crown Copyright

Bald brood is not a disease: Bald brood is not a bee disease caused by bacteria, viruses, parasites or fungi.

Symptom of wax moth infestation: It arises due to the destructive actions of wax moth larvae (Galleria mellonella).

Wax moth lifecycle: Wax moth females lay eggs near the hive or on frames. Hatching larvae tunnel through the comb, feeding on beeswax, honey, and pollen. As they grow, they damage and break through pupae cells, leading to bald brood.

Transmission:

Adult moth entry: Wax moths can enter hives through cracks, gaps, or weak spots, drawn by the scent of honey and wax.

Contaminated equipment: Used frames, boxes, or tools harboring moth eggs or larvae can introduce them to new hives.

Weak colonies: Strong colonies with good hygiene practices are less susceptible to moth infestations.

Symptoms of Bald Brood:

- ▸ Uncapped pupae: Localised patches of uncapped pupae with exposed, often discoloured bodies.
- ▸ Irregular patterns: Unlike hygienic behaviour where uncapping is uniform, bald brood shows irregular and messy uncapping patterns.
- ▸ Wax debris: Presence of spilled honey, wax fragments, webbing, and moth faeces around affected areas.
- ▸ Weakened colony: Reduced bee activity, increased drone production, and overall colony decline can accompany bald brood.

Identification:

- ▸ Visual inspection: Observe the brood combs for uncapped pupae, irregular patterns, and associated debris.
- ▸ Microscopic examination: Examining larvae or pupae under a microscope can reveal chewing marks or moth presence.
- ▸ Moth traps: Placing pheromone traps near the hive can help detect and monitor moth activity.

Impact on Colonies:

Bald Brood
Bald Brood on a frame with developing pupa clearly visible

Courtesy the Animal and Plant Health Agency © Crown Copyright

- ▸ Reduced brood production: Loss of pupae and weakened colony focus on defending against moths impact brood rearing.
- ▸ Honey and pollen loss: Wax moth larvae consume both honey and pollen, reducing available resources for the colony.
- ▸ Hive damage: Tunnelling larvae weaken and damage combs, increasing structural instability.
- ▸ Colony collapse: Severe infestations can lead to colony decline and potential collapse.

Mitigation Strategies:

Preventative measures:

- ▸ Strong colony management: Encourage strong, hygienic colonies through adequate nutrition, proper ventilation, and queen health monitoring.
- ▸ Hive maintenance: Regularly inspect hives for moth signs, repair weaknesses, and remove old comb.
- ▸ Moth deterrents: Use essential oils, diatomaceous earth, or other natural moth repellents around the apiary.

Monitoring:

- ▸ Regular hive inspections for early detection of moth activity and bald brood.
- ▸ Utilise moth traps to monitor populations and gauge potential infestation risks.
- ▸ Control measures:
- ▸ Chemical treatments: Frame-based acaricides with moth-killing properties can be used in severe cases.

▸ Freezing frames: Placing infested frames in a freezer for 24-48 hours can kill moth larvae and eggs.
▸ Cutting out affected areas: Carefully remove and destroy heavily infested comb sections.

20. BEE VIRUS X

Bee virus X (BVX) is a single strand RNA virus which is classified as follows:

Classification

- Kingdom: Viruses
- Phylum: Riboviria
- Class: Picornavirales
- Order: Picornavirales
- Family: Dicistroviridae
- Genus: Cripavirus
- Species: Apis mellifera virus X

Biology

The virus is about 7.8 kilobases long and has a single open reading frame (ORF) that encodes a polyprotein with protease, helicase and RNA-dependent RNA polymerase domains, as well as a single jelly-roll structural protein domain.

BVX replicates in the cytoplasm of honeybee cells. The virus first binds to the cell surface and is then internalised. Once inside the cell, the virus's RNA is translated into a polyprotein, which is then cleaved into individual proteins. The viral proteins then work together to replicate the viral RNA and produce new viruses.

Transmission

BVX is transmitted from bee to bee through contact with infected bees or their faeces. The virus can also be transmitted through contaminated hive equipment.

Symptoms

The symptoms of BVX infection in honeybees are not always obvious. However, some common symptoms include:

- Paralysis
- Trembling
- Death

How to identify the virus

BVX infection can be identified by testing honeybee samples for the presence of the virus's RNA. This can be done using a variety of methods, such as real-time PCR (qPCR) or reverse transcription PCR (RT-PCR).

Impact on colonies

BVX infection can have a significant impact on honeybee colonies. Heavy infestation can lead to the collapse of a colony. BVX infection can also weaken the colony and make it more susceptible to other problems, such as varroa mite infestation and American foulbrood.

Mitigation

There is no cure for BVX infection in honeybees. However, there are a number of things that beekeepers can do to mitigate the impact of the disease:

- Keep the hive strong and healthy.
- Inspect the hive regularly for signs of disease.
- Avoid overcrowding the hive.
- Use clean and sanitised hive equipment.
- Avoid introducing new bees into the hive without first inspecting them for disease.
- Purchase bees from reputable beekeepers.
- Treat the hive with an approved miticide to control varroa mites.
- Provide the bees with a balanced diet.
- Monitor the hive for signs of stress, such as increased swarming or reduced honey production.

Additional advice on mitigation:

Researchers are working to develop new ways to control BVX infection. One promising approach is to develop vaccines against the virus. Another promising approach is to use RNA interference (RNAi) to silence the virus's genes

21. BEE VIRUS Y

Bee virus Y (BVY) is a single-stranded RNA virus that is classified as follows:

Classification

- ▶ Kingdom: Viruses
- ▶ Phylum: Riboviria
- ▶ Class: Picornavirales
- ▶ Order: Picornavirales
- ▶ Family: Dicistroviridae
- ▶ Genus: Cripavirus
- ▶ Species: Apis mellifera virus Y

BVY is closely related to Bee virus X (BVX) and the two viruses are often found together in honeybee colonies. However, BVY has a number of unique features, including a different genome sequence and a different replication cycle.

Biology

BVY is a relatively small virus, with a genome that is about 7.8 kilobases long. The virus replicates in the cytoplasm of honeybee cells. The first step in the replication cycle is the attachment of the virus to the cell surface. The virus then enters the cell and its RNA is translated into a polyprotein. The polyprotein is then cleaved into individual proteins, including a protease, a helicase and an RNA-dependent RNA polymerase. These proteins work together to replicate the viral RNA and produce new viruses.

Transmission

BVY is transmitted from bee to bee through contact with infected bees or their faeces. The virus can also be transmitted through contaminated hive equipment.

Symptoms

The symptoms of BVY infection in honeybees are not always obvious. However, some common symptoms include:

- ▶ Paralysis
- ▶ Trembling
- ▶ Death

BVY can also shorten the lifespan of honeybees and exacerbate the pathogenicity of Nosema spp.

How to identify the virus

BVY infection can be identified by testing honeybee samples for the presence of the virus's RNA. This can be done using a variety of methods, such as real-time PCR (qPCR) or reverse transcription PCR (RT-PCR).

Impact on colonies

BVY infection can have a significant impact on honeybee colonies. Heavy infestation can lead to the collapse of a colony. BVY infection can also weaken the colony and make it more susceptible to other problems, such as varroa mite infestation and American foulbrood.

Mitigation

There is no cure for BVY infection in honeybees. However, there are a number of things that beekeepers can do to mitigate the impact of the disease:

- Keep the hive strong and healthy.
- Inspect the hive regularly for signs of disease.
- Avoid overcrowding the hive.
- Use clean and sanitised hive equipment.
- Avoid introducing new bees into the hive without first inspecting them for disease.
- Purchase bees from reputable beekeepers.
- Treat the hive with an approved miticide to control varroa mites.
- Provide the bees with a balanced diet.
- Monitor the hive for signs of stress, such as increased swarming or reduced honey production.

Additional advice on mitigation:

Researchers are working to develop new ways to control BVY infection. One promising approach is to develop vaccines against the virus. Another promising approach is to use RNA interference (RNAi) to silence the virus's genes.

22. BLACK QUEEN BEE CELL VIRUS

Black queen cell virus (BQCV) is a single-stranded RNA virus that is classified as follows:

Classification

▶ Kingdom: Viruses
▶ Phylum: Riboviria
▶ Class: Picornavirales
▶ Order: Picornavirales
▶ Family: Dicistroviridae
▶ Genus: Triatovirus
▶ Species: Black bee cell virus

BQCV is one of the most common and widespread honeybee viruses. It is found in honeybee colonies all over the world.

Biology

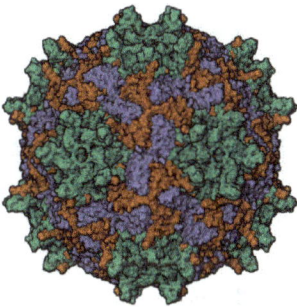

5MQC Structure of black queen cell virus

COURTESY RCSB.ORG
PDB DOI: https://doi.org/10.2210/pdb5MQC/pdb
Classification: VIRUS
Organism(s): Black queen cell virus
Deposition Author(s):
Spurny, R., Kiem, H.H.T., Plevka, P.

BQCV is a relatively small virus, with a genome that is about 8.5 kilobases long. The virus replicates in the cytoplasm of honeybee cells. The first step in the replication cycle is the attachment of the virus to the cell surface. The virus then enters the cell and its RNA is translated into a polyprotein. The polyprotein is then cleaved into individual proteins, including a protease, a helicase and an RNA-dependent RNA polymerase. These proteins work together to replicate the viral RNA and produce new viruses.

Transmission

BQCV is transmitted from bee to bee through contact with infected bees or their faeces. The virus can also be transmitted through contaminated hive equipment.

Symptoms

The symptoms of BQCV infection in honeybees include:

▶ Discoloured larvae
▶ Dead larvae
▶ Deformed bees
▶ Weakened colonies

▶ Increased susceptibility to other diseases

How to identify the virus

BQCV infection can be identified by testing honeybee samples for the presence of the virus's RNA. This can be done using a variety of methods, such as real-time PCR (qPCR) or reverse transcription PCR (RT-PCR).

Impact on colonies

BQCV infection can have a significant impact on honeybee colonies. Heavy infestation can lead to the collapse of a colony. BQCV infection can also weaken the colony and make it more susceptible to other problems, such as varroa mite infestation and American foulbrood.

Mitigation

There is no cure for BQCV infection in honeybees. However, there are a number of things that beekeepers can do to mitigate the impact of the disease:

▶ Keep the hive strong and healthy.
▶ Inspect the hive regularly for signs of disease.
▶ Avoid overcrowding the hive.
▶ Use clean and sanitised hive equipment.
▶ Avoid introducing new bees into the hive without first inspecting them for disease.
▶ Purchase bees from reputable beekeepers.
▶ Treat the hive with an approved miticide to control varroa mites.
▶ Provide the bees with a balanced diet.
▶ Monitor the hive for signs of stress, such as increased swarming or reduced honey production.

Additional advice on mitigation:

Researchers are working to develop new ways to control BQCV infection. One promising approach is to develop vaccines against the virus. Another promising approach is to use RNA interference (RNAi) to silence the virus's genes.

23. BLACK QUEEN CELL VIRUS-LIKE VIRUS

Black queen cell virus-like virus (BQCVLV) is a relatively new virus that was first identified in honeybees in Australia in 2010. It is a member of the family Dicistroviridae, which also includes other viruses that affect honeybees, such as black queen cell virus (BQCV) and acute bee paralysis virus (ABPV).

Classification

- Kingdom: Riboviria
- Phylum: Pisuviricotina
- Order: Picornavirales
- Family: Dicistroviridae
- Genus: Cripavirus
- Species: Black bee cell virus-like virus

Biology

BQCVLV is a single-stranded RNA virus with a genome that is approximately 7,500 nucleotides long. The virus replicates in the cytoplasm of infected cells and produces two capsid proteins. The virus particles are approximately 30 nanometers in diameter.

Transmission

BQCVLV is transmitted to honeybees through contact with contaminated food or water. The virus can also be transmitted through contact with contaminated hive equipment.

Symptoms

BQCVLV infection in honeybees can be asymptomatic. However, in heavily infected colonies, the virus can cause paralysis and death in adult bees. The virus can also cause queen larvae to die, resulting in black queen cells.

How to identify BQCVLV

BQCVLV can be identified by laboratory testing of honeybee samples. The most common methods of detection include:

- Reverse transcription-polymerase chain reaction (RT-PCR): This test can detect the presence of BQCVLV RNA in honeybee samples.
- Ensyme-linked immunosorbent assay (ELISA): This test can detect the presence of BQCVLV antibodies in honeybee samples.

Impact on colonies

BQCVLV infection can have a significant impact on honeybee colonies. Heavy infection can lead to:

- Reduced honey production
- Increased mortality
- Queen failure

▸ Colony collapse

Advice on mitigation

There is no cure for BQCVLV infection in honeybees. However, there are a number of things that beekeepers can do to mitigate the impact of the infection:

▸ Maintain strong colony health
▸ Avoid contact with contaminated materials
▸ Clean and sanitise hive equipment regularly
▸ Monitor colony health regularly
▸ Isolate infected colonies
▸ Destroy infected hives
▸ Purchase bees from reputable beekeepers

Additional advice

Educate yourself and others about BQCVLV and how to identify it.
Support research into the biology and control of BQCVLV.

24. BOTRYTIS CINEREA

Botrytis cinerea is a cosmopolitan fungus that is classified as follows:

Classification

- Kingdom: Fungi
- Phylum: Ascomycota
- Class: Leotiomycetes
- Order: Helotiales
- Family: Sclerotiniaceae
- Genus: Botrytis
- Species: Botrytis cinerea

Biology

Botrytis cinerea is a cosmopolitan fungus that is found in soil, air and plant material. It is a common plant pathogen that can cause a variety of diseases, including grey mold, on a wide range of crops.

Botrytis cinerea can also infect honeybees. The fungus can enter the hive through contaminated food, pollen, or hive equipment. Once inside the hive, Botrytis cinerea can infect the bees' gut and nervous system. The fungus can also cause disease in the bees' brood.

Transmission

Botrytis cinerea is transmitted from bee to bee through contact with infected bees or their faeces. The fungus can also be transmitted through contaminated food, pollen, or hive equipment.

Symptoms

The symptoms of Botrytis cinerea infection in honeybees include:

- Discoloured brood
- Dead brood
- Deformed bees
- Weakened colonies
- Increased susceptibility to other diseases

How to identify the fungus

Botrytis cinerea can be identified by microscopic examination of samples from the hive, such as brood, food and hive equipment. The fungus can also be identified by culturing samples on specialised media.

Impact on colonies

Botrytis cinerea infection can have a significant impact on honeybee colonies. Heavy infestation can lead to the collapse of a colony. Botrytis cinerea infection can also weaken the colony and make it more susceptible to other problems, such as varroa mite infestation and American foulbrood.

Mitigation

There is no cure for Botrytis cinerea infection in honeybees. However, there are a number of things that beekeepers can do to mitigate the impact of the disease:
Keep the hive strong and healthy.

- ▶ Inspect the hive regularly for signs of disease.
- ▶ Avoid overcrowding the hive.
- ▶ Use clean and sanitised hive equipment.
- ▶ Avoid introducing new bees into the hive without first inspecting them for disease.
- ▶ Purchase bees from reputable beekeepers.
- ▶ Treat the hive with an approved fungicide to control Botrytis cinerea.
- ▶ Provide the bees with a balanced diet.
- ▶ Monitor the hive for signs of stress, such as increased swarming or reduced honey production.

Additional advice on mitigation:

Researchers are working to develop new ways to control Botrytis cinerea infection in honeybees. One promising approach is to develop biological control agents, such as predatory fungi or bacteria. Another promising approach is to use RNA interference (RNAi) to silence the fungus's genes.

It is important to note that Botrytis cinerea is a complex fungus and there is still much that we do not know about it. However, by following the mitigation advice above, beekeepers can help to protect their colonies from the devastating effects of Botrytis cinerea infection.

25. CALICIVIRUS

Caliciviruses are non-enveloped RNA viruses which are classified as follows

Classification

- Kingdom: Viruses
- Realm: Riboviria
- Phylum: Negarnaviricota
- Class: Picornaviricetes
- Order: Picornavirales
- Family: Caliciviridae
- Genus: Norovirus

Biology

Caliciviruses are non-enveloped RNA viruses that replicate in the cytoplasm of infected cells. They have a wide host range, including humans, animals and birds. Honeybees are susceptible to several types of caliciviruses, including Norovirus mellifera (NVm).

NVm replicates in the cytoplasm of infected honeybee larvae, causing damage to the cells and eventually killing them. The virus is transmitted horizontally from bee to bee through contact with contaminated food, water, or hive equipment. It can also be transmitted by Varroa mites, which are parasitic mites that feed on honeybees.

Transmission

NVm is primarily transmitted through contact with contaminated food or water. Infected adult bees can shed the virus through their faeces or saliva, which can contaminate food and water sources for larvae. NVm can also be transmitted through contact with contaminated hive equipment, such as frames, combs and honey pots. Varroa mites can also transmit NVm to honeybee larvae by feeding on infected larvae and then transmitting the virus to healthy larvae.

Symptoms

The symptoms of NVm infection in honeybee larvae can vary depending on the strain of virus involved and the severity of the infection. Some of the most common symptoms include:

- Reduced brood production
- Deformed larvae
- Dead larvae
- Sticky larvae

How to identify NVm

NVm can be identified by laboratory testing of honeybee samples. The most common methods of detection include:

- Reverse transcription-polymerase chain reaction (RT-PCR): This test can detect the presence of NVm RNA in honeybee samples.

▸ Ensyme-linked immunosorbent assay (ELISA): This test can detect the presence of NVm antibodies in honeybee samples.Ensyme-linked immunosorbent assay (ELISA): This test can detect the presence of NVm antibodies in honeybee samples.

Impact on colonies

NVm infection can have a significant impact on honeybee colonies. Heavy infection can lead to:

▸ Reduced honey production
▸ Increased mortality
▸ Colony collapse

Advice on mitigation

There is no cure for NVm infection in honeybees. However, there are a number of things that beekeepers can do to mitigate the impact of the virus:

▸ Maintain strong colony health: Strong colonies are better able to withstand the effects of NVm infection.
▸ Provide adequate nutrition: Ensure that the colony has access to a plentiful supply of pollen and other essential nutrients.
▸ Control hive temperature: Avoid extreme temperatures in the hive by providing adequate ventilation and insulation.
▸ Avoid pesticide exposure: Use pesticides only as a last resort and follow label directions carefully.
▸ Monitor colony health: Regularly inspect the colony for signs of NVm infection, such as reduced brood production, deformed larvae, dead larvae and sticky larvae.
▸ Isolate infected colonies: If you find NVm infection in a colony, isolate it from other colonies to prevent the spread of the virus.
▸ Destroy infected hives: If a colony is heavily infected with NVm, it is best to destroy the hive and all its contents.
▸ Purchase bees from reputable beekeepers: Purchase bees from reputable beekeepers who have taken steps to control NVm in their colonies.
▸ Monitor Varroa mite infestation: Control Varroa mite infestation as Varroa mites can transmit NVm.

26. CHALK BROOD

See ascophaeris apis

27. CHILLED BROOD

Chilled brood is not a disease but a condition that occurs when honeybee brood is exposed to cold temperatures for an extended period of time.

Biology

Honeybee brood needs to be maintained at a temperature of around 35 degrees Celsius (95 degrees Fahrenheit) in order to develop properly. If the temperature drops below this level, the brood will become chilled. Chilled brood can be caused by a number of factors, including:

▸ Cold weather
▸ A lack of insulation in the hive
▸ A shortage of bees to keep the brood warm
▸ The hive being left open for too long during an inspection

Transmission

Chilled brood cannot be transmitted from bee to bee. However, it can be spread throughout the hive if the brood is not warmed up quickly enough.

Symptoms

The symptoms of chilled brood include:

▸ Discoloured larvae
▸ Dead larvae
▸ Deformed bees

How to identify chilled brood

Chilled brood can be identified by inspecting the hive for brood that is discoloured, dead, or deformed. If the brood is cold to the touch, it is likely chilled.

Impact on colonies

Chilled brood can have a significant impact on honeybee colonies. Heavy infestation can lead to the collapse of a colony. Chilled brood can also weaken the colony and make it more susceptible to other problems, such as varroa mite infestation and American foulbrood.

Mitigation

There are a number of things that beekeepers can do to mitigate the impact of chilled brood:

▸ Insulate the hive with a layer of straw or other insulating material.
▸ Keep the hive entrance closed during cold weather.

- ▸ Inspect the hive regularly for signs of chilled brood and rewarm any chilled brood as soon as possible.
- ▸ Avoid leaving the hive open for too long during inspections.
- ▸ Keep the colony strong and healthy.

Additional advice on mitigation:

If the colony is heavily infested with chilled brood, it may be necessary to requeen the colony.

Researchers are working to develop new ways to control chilled brood. One promising approach is to develop bees that are more resistant to cold temperatures.

28. CHILO IRIDESCENT VIRUS

Chilo iridescent virus is an enveloped, double-stranded DNA virus which is classified as follows

Classification

- ▶ Kingdom: Virus
- ▶ Realm: Riboviridae
- ▶ Phylum: Negarnaviricota
- ▶ Class: DNA
- ▶ Order: Picornavirales
- ▶ Family: Iridoviridae
- ▶ Genus: IridoviruS
- ▶ Species: Chilo iridescent

Biology

CIV is an enveloped, double-stranded DNA virus that replicates in the nucleus of infected cells. It has a wide host range, including humans, animals and insects. Honeybees are susceptible to CIV, which is also known as iridescent virus.

CIV replicates in the nucleus of infected honeybee cells, causing damage to the cells and eventually killing them. The virus is transmitted horizontally from bee to bee through contact with contaminated food, water, or hive equipment. It can also be transmitted by Varroa mites, which are parasitic mites that feed on honeybees.

Transmission

CIV is primarily transmitted through contact with contaminated food or water. Infected adult bees can shed the virus through their faeces or saliva, which can contaminate food and water sources for other bees. CIV can also be transmitted through contact with contaminated hive equipment, such as frames, combs and honey pots. Varroa mites can also transmit CIV to honeybee larvae by feeding on infected larvae and then transmitting the virus to healthy larvae.

Symptoms

The symptoms of CIV infection in honeybees can vary depending on the strain of virus involved and the severity of the infection. Some of the most common symptoms include:

- ▶ Reduced honey production
- ▶ Increased mortality
- ▶ Deformed wings
- ▶ Paralysis
- ▶ Hair loss
- ▶ Darkened abdomens
- ▶ Sticky larvae

  ▶   Iridescent colouration of the abdomen

Identification

CIV can be identified by laboratory testing of honeybee samples. The most common methods of detection include:

  ▶ Electron microscopy: This method can visualise the virus particles in infected larvae.
  ▶ Polymerase chain reaction (PCR): This test can detect the presence of CIV DNA in honeybee samples.
  ▶ Ensyme-linked immunosorbent assay (ELISA): This test can detect the presence of CIV antibodies in honeybee samples.

Impact on colonies

CIV infection can have a significant impact on honeybee colonies. Heavy infection can lead to:

  ▶   Reduced honey production
  ▶   Increased mortality
  ▶   Colony collapse

Advice on mitigation

There is no cure for CIV infection in honeybees. However, there are a number of things that beekeepers can do to mitigate the impact of the virus:

  ▶   Maintain strong colony health: Strong colonies are better able to withstand the effects of CIV infection.
  ▶   Provide adequate nutrition: Ensure that the colony has access to a plentiful supply of pollen and other essential nutrients.
  ▶   Control hive temperature: Avoid extreme temperatures in the hive by providing adequate ventilation and insulation.
  ▶   Avoid pesticide exposure: Use pesticides only as a last resort and follow label directions carefully.
  ▶   Monitor colony health: Regularly inspect the colony for signs of CIV infection, such as reduced honey production, increased mortality, deformed wings, paralysis, hair loss, darkened abdomens, sticky larvae and iridescent colouration of the abdomen.
  ▶   Isolate infected colonies: If you find CIV infection in a colony, isolate it from other colonies to prevent the spread of the virus.
  ▶   Destroy infected hives: If a colony is heavily infected with CIV, it is best to destroy the hive and all its contents.
  ▶   Purchase bees from reputable beekeepers: Purchase bees from reputable beekeepers who have taken steps to control CIV in their colonies.
  ▶   Monitor Varroa mite infestation: Control Varroa mite infestation as Varroa mites can transmit CIV.

29. CHRONIC BEE PARALYSIS VIRUS

Chronic bee paralysis virus (CBPV) is a single-stranded RNA virus that is classified as follows:

Classification

- Kingdom: Viruses
- Phylum: Riboviria
- Class: Picornavirales
- Order: Picornavirales
- Family: Dicistroviridae
- Genus: Cripavirus
- Species: Chronic bee paralysis virus

Biology

CBPV is a relatively small virus, with a genome that is about 7.8 kilobases long. The virus replicates in the cytoplasm of honeybee cells. The first step in the replication cycle is the attachment of the virus to the cell surface. The virus then enters the cell and its RNA is translated into a polyprotein. The polyprotein is then cleaved into individual proteins, including a protease, a helicase and an RNA-dependent RNA polymerase. These proteins work together to replicate the viral RNA and produce new viruses.

Transmission

CBPV is transmitted from bee to bee through contact with infected bees or their faeces. The virus can also be transmitted through contaminated hive equipment.

Symptoms

The symptoms of CBPV infection in honeybees include:
- Ataxia (incoordination)
- Trembling
- Paralysis
- Hair loss
- Shortened abdomen
- Black colouration

How to identify the virus

CBPV infection can be identified by testing honeybee samples for the presence of the virus's RNA. This can be done using a variety of methods, such as real-time PCR (qPCR) or reverse transcription PCR (RT-PCR).

Impact on colonies

CBPV infection can have a significant impact on honeybee colonies. Heavy infestation can lead to the collapse of a colony. CBPV infection can also weaken the colony and make it more susceptible to other problems, such as varroa mite infestation and American foulbrood.

Mitigation

There is no cure for CBPV infection in honeybees. However, there are a number of things that beekeepers can do to mitigate the impact of the disease:

▸ Keep the hive strong and healthy.
▸ Inspect the hive regularly for signs of disease.
▸ Avoid overcrowding the hive.
▸ Use clean and sanitised hive equipment.
▸ Avoid introducing new bees into the hive without first inspecting them for disease.
▸ Purchase bees from reputable beekeepers.
▸ Treat the hive with an approved miticide to control varroa mites.
▸ Provide the bees with a balanced diet.
▸ Monitor the hive for signs of stress, such as increased swarming or reduced honey production.

By following this advice, beekeepers can help to protect their colonies from CBPV infection and prevent the spread of the virus.

Additional advice on mitigation:

Researchers are working to develop new ways to control CBPV infection. One promising approach is to develop vaccines against the virus. Another promising approach is to use RNA interference (RNAi) to silence the virus's genes.

30. CHRONIC BEE PARALYSIS VIRUS-LIKE VIRUS

Chronic bee paralysis virus-like virus (CBPV-LV) is a single-stranded RNA virus that is closely related to Chronic bee paralysis virus (CBPV). However, CBPV-LV is not formally classified as a separate virus species.

Classification

CBPV-LV is classified as follows:

- Kingdom: Viruses
- Phylum: Riboviria
- Class: Picornavirales
- Order: Picornavirales
- Family: Dicistroviridae
- Genus: Cripavirus
- Species: Chronic bee paralysis virus-like virus

Biology

CBPV-LV is a relatively small virus, with a genome that is about 7.8 kilobases long. The virus replicates in the cytoplasm of honeybee cells. The first step in the replication cycle is the attachment of the virus to the cell surface. The virus then enters the cell and its RNA is translated into a polyprotein. The polyprotein is then cleaved into individual proteins, including a protease, a helicase and an RNA-dependent RNA polymerase. These proteins work together to replicate the viral RNA and produce new viruses.

Transmission

CBPV-LV is transmitted from bee to bee through contact with infected bees or their faeces. The virus can also be transmitted through contaminated hive equipment.

Symptoms

The symptoms of CBPV-LV infection in honeybees are similar to those of CBPV infection and include:

- Ataxia (incoordination)
- Trembling
- Paralysis
- Hair loss
- Shortened abdomen
- Black colouration

However, CBPV-LV is generally considered to be less virulent than CBPV.

How to identify the virus

CBPV-LV infection can be identified by testing honeybee samples for the presence of the virus's RNA. This can be done using a variety of methods, such as real-time PCR (qPCR) or reverse transcription PCR (RT-PCR).

Impact on colonies

CBPV-LLV infection can have a significant impact on honeybee colonies. Heavy infestation can lead to the collapse of a colony. CBPV-LV infection can also weaken the colony and make it more susceptible to other problems, such as varroa mite infestation and American foulbrood.

Mitigation

There is no cure for CBPV-LV infection in honeybees. However, there are a number of things that beekeepers can do to mitigate the impact of the disease:

▸ Keep the hive strong and healthy.
▸ Inspect the hive regularly for signs of disease.
▸ Avoid overcrowding the hive.
▸ Use clean and sanitised hive equipment.
▸ Avoid introducing new bees into the hive without first inspecting them for disease.
▸ Purchase bees from reputable beekeepers.
▸ Treat the hive with an approved miticide to control varroa mites.
▸ Provide the bees with a balanced diet.
▸ Monitor the hive for signs of stress, such as increased swarming or reduced honey production.

Additional advice on mitigation:

Researchers are working to develop new ways to control CBPV-LV infection. One promising approach is to develop vaccines against the virus. Another promising approach is to use RNA interference (RNAi) to silence the virus's genes.

31. CIRCOVIRUS

Circovirus are single strand DNA viruses which are classified as follows

Classification

- ▶ Kingdom: Viruses
- ▶ Realm: Riboviria
- ▶ Phylum: Negarnaviricota
- ▶ Class: Geminiviricetes
- ▶ Order: Circovirales
- ▶ Family: Circoviridae
- ▶ Genus: Circovirus
- ▶ Species: Circovirus

Biology

Circoviruses are single-stranded DNA viruses that replicate in the nucleus of infected cells. They have a wide host range, including humans, animals and insects. Honeybees are susceptible to several types of circoviruses, including Honeybee circovirus (HBcV) and Beekeeper circovirus (BvCV).

HBcV and BvCV replicate in the nucleus of infected honeybee cells, causing damage to the cells and eventually killing them. The viruses are transmitted horizontally from bee to bee through contact with contaminated food, water, or hive equipment.

Transmission

HBcV and BvCV are primarily transmitted through contact with contaminated food or water. Infected adult bees can shed the viruses through their faeces or saliva, which can contaminate food and water sources for other bees. HBcV and BvCV can also be transmitted through contact with contaminated hive equipment, such as frames, combs and honey pots.

Symptoms

The symptoms of Circoviridae infection in honeybees can vary depending on the strain of virus involved and the severity of the infection. Some of the most common symptoms include:

- ▶ Reduced honey production
- ▶ Increased mortality
- ▶ Deformed wings
- ▶ Paralysis
- ▶ Hair loss
- ▶ Darkened abdomens

How to identify Circoviridae

Circoviridae can be identified by laboratory testing of honeybee samples. The most common methods of detection include:

▸ Polymerase chain reaction (PCR): This test can detect the presence of Circoviridae DNA in honeybee samples.

▸ Ensyme-linked immunosorbent assay (ELISA): This test can detect the presence of Circoviridae antibodies in honeybee samples.

Impact on colonies

Circoviridae infection can have a significant impact on honeybee colonies. Heavy infection can lead to:

▸ Reduced honey production

▸ Increased mortality

▸ Colony collapse

Advice on mitigation

There is no cure for Circoviridae infection in honeybees. However, there are a number of things that beekeepers can do to mitigate the impact of the viruses:

▸ Maintain strong colony health: Strong colonies are better able to withstand the effects of Circoviridae infection.

▸ Provide adequate nutrition: Ensure that the colony has access to a plentiful supply of pollen and other essential nutrients.

▸ Control hive temperature: Avoid extreme temperatures in the hive by providing adequate ventilation and insulation.

▸ Avoid pesticide exposure: Use pesticides only as a last resort and follow label directions carefully.

▸ Monitor colony health: Regularly inspect the colony for signs of Circoviridae infection, such as reduced honey production, increased mortality, deformed wings, paralysis, hair loss and darkened abdomens.

▸ Isolate infected colonies: If you find Circoviridae infection in a colony, isolate it from other colonies to prevent the spread of the viruses.

▸ Destroy infected hives: If a colony is heavily infected with Circoviridae, it is best to destroy the hive and all its contents.

▸ Purchase bees from reputable beekeepers: Purchase bees from reputable beekeepers who have taken steps to control Circoviridae in their colonies.

32. CLADOSPORIUM CLADOSPORIOIDES

Cladosporium Cladosporioides is a saprobic fungal disease which is classified as follows

Classification

- Kingdom: Fungi
- Phylum: Ascomycota
- Class: Dothideomycetes
- Order: Capnondiales
- Family: Cladosporiaceae
- Genus: Cladosporium
- Species: Cladosporium cladosporioides

Biology

Cladosporium cladosporioides is a saprobic fungus that is commonly found in soil, water and air. It is one of the most common airborne fungi in the world. Honeybees can be exposed to Cladosporium cladosporioides through contact with contaminated pollen, water, or hive equipment. The fungus can also be transmitted from bee to bee through contact with contaminated faeces or saliva.

Once a honeybee is infected with Cladosporium cladosporioides, the fungus can grow in the bee's gut and produce toxins that can damage the bee's cells. The fungus can also spread to other parts of the bee's body, such as the thorax and abdomen.

Transmission

Cladosporium cladosporioides is primarily transmitted through contact with contaminated food or water. Infected adult bees can shed the fungus through their faeces or saliva, which can contaminate food and water sources for other bees. Cladosporium cladosporioides can also be transmitted through contact with contaminated hive equipment, such as frames, combs and honey pots.

Symptoms

The symptoms of Cladosporium cladosporioides infection in honeybees can vary depending on the severity of the infection. Some of the most common symptoms include:

- Reduced honey production
- Increased mortality
- Deformed wings
- Paralysis
- Hair loss
- Darkened abdomens
- Sticky larvae

Identification

Cladosporium cladosporioides can be identified by laboratory testing of honeybee samples. The most common methods of detection include:

▶ Microscopic examination: This method can visualise the fungus in infected bee faeces or larvae.

▶ Culture: This method can grow the fungus in a laboratory setting, which allows for further identification.

▶ Polymerase chain reaction (PCR): This test can detect the presence of Cladosporium cladosporioides DNA in honeybee samples.

Impact on colonies

Cladosporium cladosporioides infection can have a significant impact on honeybee colonies. Heavy infection can lead to:

▶ Reduced honey production

▶ Increased mortality

▶ Colony collapse

Advice on mitigation

There is no cure for Cladosporium cladosporioides infection in honeybees. However, there are a number of things that beekeepers can do to mitigate the impact of the fungus:

▶ Maintain strong colony health: Strong colonies are better able to withstand the effects of Cladosporium cladosporioides infection.

▶ Provide adequate nutrition: Ensure that the colony has access to a plentiful supply of pollen and other essential nutrients.

▶ Control hive temperature: Avoid extreme temperatures in the hive by providing adequate ventilation and insulation.

▶ Avoid pesticide exposure: Use pesticides only as a last resort and follow label directions carefully.

▶ Monitor colony health: Regularly inspect the colony for signs of Cladosporium cladosporioides infection, such as reduced honey production, increased mortality, deformed wings, paralysis, hair loss, darkened abdomens and sticky larvae.

▶ Isolate infected colonies: If you find Cladosporium cladosporioides infection in a colony, isolate it from other colonies to prevent the spread of the fungus.

▶ Destroy infected hives: If a colony is heavily infected with Cladosporium cladosporioides, it is best to destroy the hive and all its contents.

▶ Purchase bees from reputable beekeepers: Purchase bees from reputable beekeepers who have taken steps to control Cladosporium cladosporioides in their colonies.

▶ Monitor Varroa mite infestation: Control Varroa mite infestation as Varroa mites can transmit Cladosporium cladosporioides.

33. CLOUDY WING VIRUS

Cloudy wing virus is a relatively small icosahedral virus which is classified as follows

Classification

- Kingdom: Viruses
- Phylum: Riboviria
- Class: Picornavirales
- Order: Picornavirales
- Family: Iflaviridae
- Genus: Iflavirus
- Species: Cloudy wing virus

Biology

CWV is a relatively small virus, with a genome that is about 10.5 kilobases long. The virus replicates in the cytoplasm of honeybee cells. The first step in the replication cycle is the attachment of the virus to the cell surface. The virus then enters the cell and its RNA is translated into a polyprotein. The polyprotein is then cleaved into individual proteins, including a protease, a helicase and an RNA-dependent RNA polymerase. These proteins work together to replicate the viral RNA and produce new viruses.

Transmission

CWV is transmitted from bee to bee through contact with infected bees or their faeces. The virus can also be transmitted through contaminated hive equipment.

Symptoms

The symptoms of CWV infection in honeybees include:

- Cloudy wings
- Deformed wings
- Discoloured wings
- Shortened lifespan
- Weakened colony
- Increased susceptibility to other diseases

How to identify the virus

CWV infection can be identified by testing honeybee samples for the presence of the virus's RNA. This can be done using a variety of methods, such as real-time PCR (qPCR) or reverse transcription PCR (RT-PCR).

Impact on colonies

CWV infection can have a significant impact on honeybee colonies. Heavy infestation can lead to the collapse of a colony. CWV infection can also weaken the colony and make it more susceptible to other problems, such as varroa mite infestation and

American foulbrood.

Mitigation

There is no cure for CWV infection in honeybees. However, there are a number of things that beekeepers can do to mitigate the impact of the disease:

- Keep the hive strong and healthy.
- Inspect the hive regularly for signs of disease.
- Avoid overcrowding the hive.
- Use clean and sanitised hive equipment.
- Avoid introducing new bees into the hive without first inspecting them for disease.
- Purchase bees from reputable beekeepers.
- Treat the hive with an approved miticide to control varroa mites.
- Provide the bees with a balanced diet.
- Monitor the hive for signs of stress, such as increased swarming or reduced honey production.

Additional advice on mitigation:

Researchers are working to develop new ways to control CWV infection. One promising approach is to develop vaccines against the virus. Another promising approach is to use RNA interference (RNAi) to silence the virus's genes.

34. COLONY COLLAPSE DISORDER

Colony Collapse Disorder (CCD) is not a single disease, but rather a complex condition that is characterised by the rapid disappearance of adult worker bees from a colony. CCD is classified as a syndrome.

Biology

The exact cause of CCD is unknown, but it is thought to be caused by a combination of factors, including:

▶ Parasites and diseases, such as varroa mites and Nosema ceranae
▶ Poor nutrition
▶ Pesticides
▶ Loss of habitat
▶ Climate change

Transmission

The transmission of CCD is not fully understood, but it is thought to be caused by a combination of factors, including:

▶ Contact with infected bees
▶ Contaminated hive equipment
▶ Pollen and nectar from contaminated plants

Symptoms

The symptoms of CCD include:

▶ The sudden disappearance of adult worker bees from the hive
▶ The presence of a queen and brood in the hive
▶ The presence of food stores in the hive

How to identify CCD

CCD can be identified by the following criteria:

▶ The colony is strong and healthy in the spring, but the adult worker bees suddenly disappear in the summer or autumn.
▶ The queen and brood are still present in the hive.
▶ There are food stores present in the hive.
▶ There are no dead bees found inside or outside of the hive.

Impact on colonies

CCD can have a devastating impact on honeybee colonies. Colonies that are affected by CCD often collapse completely, leaving behind the queen and brood. This can lead to significant losses for beekeepers and farmers and it can also have a negative impact on the environment.

Advice on mitigation

There is no cure for CCD, but there are a number of things that beekeepers can do to mitigate its impact:

- ▶ Keep the hive strong and healthy.
- ▶ Inspect the hive regularly for signs of disease and pests.
- ▶ Provide the bees with a balanced diet.
- ▶ Avoid using pesticides that are harmful to bees.
- ▶ Plant bee-friendly flowers and shrubs.

Additional advice on mitigation

Researchers are working to develop new ways to control CCD. One promising approach is to develop vaccines against the parasites and diseases that are thought to contribute to CCD. Another promising approach is to develop bee breeds that are more resistant to CCD.

35. CORONAVIRUS (HONEYBEE CORONAVIRUS)

Coronaviruses are enveloped RNA viruses which are classified as follows

Classification

- Kingdom:Viruses
- Realm: Riboviridae
- Phylum: Negarnaviricota
- Class: Pisuviricetes
- Order: Nidovirales
- Family: Coronaviridae
- Genus: Alphacoronavirus

Biology

Coronaviruses are enveloped RNA viruses that replicate in the cytoplasm of infected cells. They have a wide host range, including humans, animals and birds. Honeybees are susceptible to several types of coronaviruses, including Honeybee coronavirus (HBCoV) and Israel acute paralysis virus (IAPV).

HBCoV and IAPV replicate in the cytoplasm of infected honeybee cells, causing damage to the cells and eventually killing them. The viruses are transmitted horizontally from bee to bee through contact with contaminated food, water, or hive equipment.

Transmission

HBCoV and IAPV are primarily transmitted through contact with contaminated food or water. Infected adult bees can shed the viruses through their faeces or saliva, which can contaminate food and water sources for other bees. HBCoV and IAPV can also be transmitted through contact with contaminated hive equipment, such as frames, combs and honey pots.

Symptoms

The symptoms of Coronaviridae infection in honeybees can vary depending on the strain of virus involved and the severity of the infection. Some of the most common symptoms include:

- Reduced honey production
- Increased mortality
- Deformed wings
- Paralysis
- Hair loss
- Darkened abdomens
- Tremors

How to identify Coronaviridae

Coronaviridae can be identified by laboratory testing of honeybee samples. The most common methods of detection include:

▶ Reverse transcription-polymerase chain reaction (RT-PCR): This test can detect the presence of Coronaviridae RNA in honeybee samples.

▶ Ensyme-linked immunosorbent assay (ELISA): This test can detect the presence of Coronaviridae antibodies in honeybee samples.

Impact on colonies

Coronaviridae infection can have a significant impact on honeybee colonies. Heavy infection can lead to:

▶ Reduced honey production

▶ Increased mortality

▶ Colony collapse

Advice on mitigation

There is no cure for Coronaviridae infection in honeybees. However, there are a number of things that beekeepers can do to mitigate the impact of the viruses:

▶ Maintain strong colony health: Strong colonies are better able to withstand the effects of Coronaviridae infection.

▶ Provide adequate nutrition: Ensure that the colony has access to a plentiful supply of pollen and other essential nutrients.

▶ Control hive temperature: Avoid extreme temperatures in the hive by providing adequate ventilation and insulation.

▶ Avoid pesticide exposure: Use pesticides only as a last resort and follow label directions carefully.

▶ Monitor colony health: Regularly inspect the colony for signs of Coronaviridae infection, such as reduced honey production, increased mortality, deformed wings, paralysis, hair loss, darkened abdomens and tremors.

▶ Isolate infected colonies: If you find Coronaviridae infection in a colony, isolate it from other colonies to prevent the spread of the viruses.

▶ Destroy infected hives: If a colony is heavily infected with Coronaviridae, it is best to destroy the hive and all its contents.

▶ Purchase bees from reputable beekeepers: Purchase bees from reputable beekeepers who have taken steps to control Coronaviridae in their colonies.

36. CRICKET PARALYSIS VIRUS

Cricket paralysis virus (CrPV) is a non-enveloped RNA virus that belongs to the family Dicistroviridae, order Picornavirales. It is a mosquito-borne virus that commonly infects crickets (Teleogryllus commodus and T. oceanicus) and Drosophila melanogaster and has also been found in honeybees (Apis mellifera).

Classification

▶ Kingdom: Viruses
▶ Realm: Riboviria
▶ Phylum: Negarnaviricota
▶ Class: Picornaviricetes
▶ Order: Picornavirales
▶ Family: Dicistroviridae
▶ Genus: Cripavirus

Biology

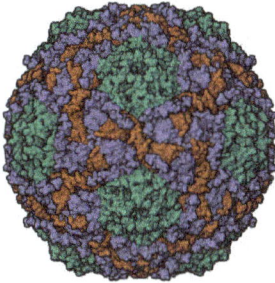

1B35 CRICKET PARALYSIS VIRUS (CRPV)
COURTESY RCSB.ORG
PDB DOI: https://doi.org/10.2210/pdb1B35/pdb
Deposition Author(s): Tate, J.G., Liljas, L., Scotti, P.D., Christian, P.D., Lin, T.W., Johnson, J.E.

CrPV replicates in the cytoplasm of infected cells. It is transmitted horizontally from insect to insect through contact with contaminated faeces or saliva. The virus can also be transmitted through contact with contaminated hive equipment.

Transmission

CrPV is primarily transmitted through contact with contaminated cricket faeces. Infected crickets can shed the virus through their faeces, which can contaminate food and water sources for other crickets and bees. CrPV can also be transmitted through contact with contaminated Drosophila melanogaster, as well as through contact with contaminated hive equipment, such as frames, combs and honey pots.

Symptoms

The symptoms of CrPV infection in honeybees can vary depending on the strain of virus involved and the severity of the infection. Some of the most common symptoms include:

▶ Paralysis: The virus can cause paralysis of the legs, wings and abdomen.
▶ Tremors: The virus can cause tremors in the legs, wings and abdomen.
▶ Death: The virus can cause death in infected bees.

How to identify CrPV

CrPV can be identified by laboratory testing of honeybee samples. The most common methods of detection include:

▶ Reverse transcription-polymerase chain reaction (RT-PCR): This test can detect the presence of CrPV RNA in honeybee samples.

▶ Ensyme-linked immunosorbent assay (ELISA): This test can detect the presence of CrPV antibodies in honeybee samples.

Impact on colonies

▶ CrPV infection can have a significant impact on honeybee colonies. Heavy infection can lead to:

▶ Reduced honey production: Infected colonies may produce less honey.

▶ Increased mortality: Infected colonies may experience increased mortality.

▶ Colony collapse: In severe cases, infected colonies may collapse completely.

Advice on mitigation

There is no cure for CrPV infection in honeybees. However, there are a number of things that beekeepers can do to mitigate the impact of the virus:

▶ Maintain strong colony health: Strong colonies are better able to withstand the effects of CrPV infection.

▶ Provide adequate nutrition: Ensure that the colony has access to a plentiful supply of pollen and other essential nutrients.

▶ Control hive temperature: Avoid extreme temperatures in the hive by providing adequate ventilation and insulation.

▶ Avoid pesticide exposure: Use pesticides only as a last resort and follow label directions carefully.

▶ Monitor colony health: Regularly inspect the colony for signs of CrPV infection, such as paralysis, tremors and death.

▶ Isolate infected colonies: If you find CrPV infection in a colony, isolate it from other colonies to prevent the spread of the virus.

▶ Destroy infected hives: If a colony is heavily infected with CrPV, it is best to destroy the hive and all its contents.

▶ Purchase bees from reputable beekeepers: Purchase bees from reputable beekeepers who have taken steps to control CrPV in their colonies.

37. DEFORMED WING VIRUS TYPE A

Deformed Wing Virus Type A (DWV-A) is a single strand negative-sense virus

Classification:

- Kingdom: Riboviria
- Phylum: Negarnaviricota
- Class: Mononegavirales
- Order: Mononegavirales
- Family: Rhabdoviridae
- Genus: Vespavirus
- Species: Deformed wing virus (DWV)

Biology:

Deformed wing virus
Worker bee showing stunted wings associated with deformed wing virus

Courtesy the Animal and Plant Health Agency © Crown Copyright

Genome: Single-stranded negative-sense RNA virus, approximately 12kb long.

Replication: Occurs inside host cells, using the host's machinery to make copies of its own RNA.

Transmission: Primarily transmitted by the parasitic mite Varroa destructor. Can also be transmitted through contact with infected bees and materials.

Host range: Primarily infects honeybees (Apis mellifera) but has been detected in other bee species and bumblebees.

Virulence: DWV-A is generally considered less virulent than other DWV strains like DWV-B, but can still significantly impact colony health.

Symptoms:

- Deformed wings: The most characteristic symptom, with wings appearing crumpled, shortened, or unable to unfold properly.
- Reduced flight ability: Infected bees struggle to fly or fly erratically.
- Tremors and paralysis: In severe cases, bees may experience tremors or even complete paralysis.
- Abdomen distention: Sometimes seen due to impaired digestion caused by the virus.
- Reduced lifespan: Infected bees typically have shorter lifespans.

Identification:

Visual inspection: Checking for deformed wings and flight ability can provide initial clues.

Polymerase chain reaction (PCR) testing: The most accurate method for confirming DWV-A infection.

Beekeeper observations: Changes in colony behaviour, such as decreased foraging activity or increased dead bees, can be suggestive of DWV infection, but further confirmation is needed.

Impact on colonies:

Reduced honey production: Infected bees contribute less to colony productivity due to flight limitations and potential energy depletion.

Increased colony mortality: Severe DWV-A infections can weaken the colony and contribute to winter losses.

Increased susceptibility to other diseases: DWV-A can compromise bee immune systems, making them more vulnerable to other pathogens.

Mitigation:

Integrated pest management (IPM): Controlling Varroa mite populations is the primary approach to preventing DWV-A transmission. This involves using a combination of chemical, biological, and cultural control methods.

Strong queen selection: Selecting healthy and productive queens can help colonies build resilience against DWV-A and other stressors.

Good beekeeping practices: Maintaining proper hive hygiene, providing adequate nutrition, and avoiding stress factors can help strengthen colony health and resistance to DWV-A.

Research and development: Ongoing research on vaccines, antiviral treatments, and other DWV-A control methods is crucial for improving bee health and colony survival.

38. DEFORMED WING VIRUS TYPE B

Deformed wing virus type B is a single strand negative-sense RNA virus.

Classification:

- ▶ Kingdom: Riboviria
- ▶ Phylum: Negarnaviricota
- ▶ Class: Mononegavirales
- ▶ Order: Mononegavirales
- ▶ Family: Rhabdoviridae
- ▶ Genus: Vespavirus
- ▶ Species: Deformed wing virus type B (DWV-B)

Biology:

Deformed Wing Virus
Honey bee displaying signs of
Deformed Wing Virus usually
associated with Varroa infestation

Courtesy the Animal and Plant
Health Agency © Crown Copyright

Genome: Single-stranded negative-sense RNA virus, approximately 12kb long.

Replication: Occurs inside host cells, using the host's machinery to make copies of its own RNA.

Transmission: Primarily transmitted by the parasitic mite Varroa destructor. Can also be transmitted through contact with infected bees and materials.

Host range: Primarily infects honeybees (Apis mellifera) but has been detected in other bee species and bumblebees.

Virulence: DWV-B is considered more virulent than DWV-A, leading to higher mortality rates in infected bees, especially during winter months.

Symptoms:

Deformed wings: Similar to DWV-A, with wings appearing crumpled, shortened, or unable to unfold properly.

Reduced flight ability: Infected bees struggle to fly or fly erratically.

Tremors and paralysis: More common than with DWV-A, and can occur even in mildly infected bees.

Abdomen distention: Sometimes seen due to impaired digestion caused by the virus.

Reduced lifespan: Significantly shorter lifespan compared to healthy bees.

Colony-level symptoms: Decreased foraging activity, increased dead bees, weakened colony.

Identification:

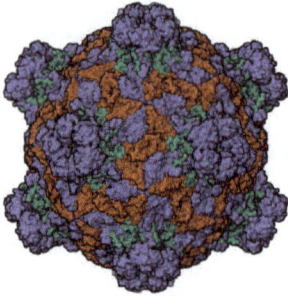

5L7Q Structure of deformed wing virus, a honeybee pathogen
COURTESY RCSB.ORG
PDB DOI: https://doi.org/10.2210/pdb5L7Q/pdb
Deposition Author(s): Skubnik, K., Novacek, J., Fuzik, T., Pridal, A., Paxton, R., Plevka, P.

Visual inspection: Checking for deformed wings and flight ability can provide initial clues. However, due to shared symptoms with DWV-A, confirmation is crucial.

Polymerase chain reaction (PCR) testing: Most accurate method for confirming DWV-B infection.

Viral load quantification: Helps assess the severity of infection and potential impact on colony health.

Beekeeper observations: Similar to DWV-A, changes in colony behaviour, but with increased emphasis on mortality and winter colony losses.

Impact on colonies:

Severe winter colony losses: DWV-B is a major contributor to overwinter colony mortality, due to its ability to replicate and reach lethal levels during the winter months when bees are confined to the hive.

Reduced colony strength and productivity: Even sublethal infections can weaken colonies, impacting honey production, pollination services, and overall colony survival.

Increased susceptibility to other diseases: DWV-B can compromise bee immune systems, making them more vulnerable to other pathogens.

Mitigation:

Integrated pest management (IPM): Controlling Varroa mite populations is the primary approach to preventing DWV-B transmission. This involves using a combination of chemical, biological, and cultural control methods.

Strong queen selection: Selecting healthy and productive queens can help colonies build resilience against DWV-B and other stressors.

Good beekeeping practices: Maintaining proper hive hygiene, providing adequate nutrition, and avoiding stress factors can help strengthen colony health and resistance to DWV-B.

Research and development: Ongoing research on vaccines, antiviral treatments, and other DWV-B control methods is crucial for improving bee health and colony survival.

39. DEFORMED WING VIRUS TYPE C

Deformed wing virus type C (DWV-C) remains a relatively elusive member of the DWV family. While not as prevalent as DWV-A and DWV-B, understanding its characteristics and potential impact is crucial for comprehensive bee health management.

Classification:

- Kingdom: Riboviria
- Phylum: Negarnaviricota
- Class: Mononegavirales
- Order: Mononegavirales
- Family: Rhabdoviridae
- Genus: Vespavirus
- Species: Deformed wing virus (DWV)

Biology:

Genome: Single-stranded negative-sense RNA virus, approximately 12kb long.

Replication: Similar to DWV-A and B, occurs inside host cells using their machinery.

Transmission: Less understood compared to other DWV types. Likely involves Varroa mites and potentially direct contact with infected bees.

Host range: Primarily honeybees (Apis mellifera) but detection in other bee species suggests broader potential.

Virulence: Remains unclear, but some studies suggest it might be less virulent than DWV-B and possibly even asymptomatic in adult bees.

Symptoms:

Limited information: Due to its elusive nature, defining specific symptoms associated with DWV-C is challenging.

Potential overlap: Deformed wings, reduced flight ability, and tremors might be present, but these can also occur with other DWV types and other bee diseases.

Colony-level effects: Possible impacts include reduced colony strength, decreased brood production, and increased winter mortality, but linking these directly to DWV-C requires further research.

Identification:

PCR testing: Currently, the only reliable method for detecting DWV-C, although it doesn't necessarily distinguish between active infection and past exposure.

Viral load quantification: Can help assess potential impact, but interpretation for DWV-C is still developing.

Beekeeper observations: Colony-level symptoms can be suggestive but require confirmation with appropriate testing.

Impact on colonies:

Uncertainty remains: The full extent of DWV-C's impact on colony health is unclear and ongoing research is needed.

Potential contribution to colony decline: Its presence in weakened colonies and association with winter mortality suggest it might play a role, but further investigation is crucial.

Complex interaction with other factors: Understanding DWV-C's role requires considering its interplay with other pathogens, Varroa mites, and environmental stressors.

Mitigation:

Focus on Varroa control: As the primary DWV vector, effective Varroa mite management remains crucial for reducing overall DWV transmission, including potential DWV-C spread.

Maintain strong colony health: Good beekeeping practices, including proper nutrition and stress reduction, can contribute to a colony's resilience against various stressors, including potential DWV-C challenges.

Support research efforts: Continued research on DWV-C's biology, transmission, and impact on bee health is vital for developing effective mitigation strategies and protecting honeybee populations.

40. DEFORMED WING VIRUS-D

Deformed wing virus-d is in the process of being how Egyptian Acute Paralysis Virus (EAPV) is reclassified.

Refer to Egyptian Acute Paralysis Virus

41. DEFORMED WING VIRUS-LIKE VIRUS

While deformed wing virus (DWV) continues to pose a significant threat to honeybees, the emergence of deformed wing virus-like viruses (DWV-like viruses) adds another layer of complexity to bee health management.

Classification:

- ▶ Kingdom: Riboviria
- ▶ Phylum: Negarnaviricota
- ▶ Class: Mononegavirales
- ▶ Order: Mononegavirales
- ▶ Family: Uncertain, potentially Rhabdoviridae, but further research is needed.
- ▶ Genus: Uncertain, with "Cripavirus" proposed as a possible candidate for both DWV and DWV-like viruses.
- ▶ Species: Deformed wing virus-like virus (temporary placeholder)

Biology:

Genome: Single-stranded negative-sense RNA, possibly around 12kb in sise, similar to known DWV strains.

Replication: Similar to DWV, likely utilises host cells' machinery to convert its RNA to a positive-sense form for replication.

Transmission: Suspected to involve Varroa mites as primary vectors, but additional routes like direct contact with infected bees might also be possible.

Host range: Primarily Apis mellifera, but investigations are ongoing for potential presence in other bee species.

Symptoms:

Limited information: Due to its recent discovery and ongoing research, definitive symptoms remain elusive.

Potential overlap with DWV: Deformed wings, reduced flight ability, tremors, and abdominal distention might occur, but these are also associated with DWV and other bee diseases.

Colony-level effects: Possible impacts include decreased brood production, reduced colony strength, and increased winter mortality, but linking these directly to DWV-like viruses requires further study.

Identification:

PCR testing: Currently, the only reliable method for detecting DWV-like viruses, although it doesn't necessarily distinguish between active infection and past exposure.

Viral load quantification: Can help assess potential impact, but interpretation for DWV-like viruses is still under development.

Beekeeper observations: While colony-level symptoms can be suggestive, confirmation through appropriate testing is crucial.

Impact on colonies:

Uncertain territory: The full extent of DWV-like virus impact on colony health remains unclear and ongoing research is essential.

Potential contribution to colony decline: Its presence in weakened colonies and possible association with winter mortality suggest it might play a role, but further investigation is needed.

Complex interplay with other factors: Understanding DWV-like virus impact requires considering its interaction with other pathogens, Varroa mites, and environmental stressors.

Mitigation:

Focus on Varroa control: As the primary DWV vector, effective Varroa mite management remains crucial for reducing overall DWV-like virus transmission, as well as protecting against other honeybee diseases.

Maintain strong colony health: Good beekeeping practices, including proper nutrition, adequate space, and stress reduction, can contribute to a colony's resilience against various stressors, potentially including DWV-like viruses.

Support research efforts: Continued research on DWV-like virus biology, transmission, and impact on colony health is vital for developing effective mitigation strategies and safeguarding honeybee populations.

42. DYSENTERY

Dysentery in honeybees is not a single disease, but rather a symptom of a number of different conditions. It is characterised by diarrhoea, which can be caused by a variety of factors, including:

Infection with parasites or diseases, such as Nosema ceranae or Bacillus alvei

- ▶ Poor diet
- ▶ Cold weather
- ▶ Stress

Biology

Dysentery occurs when the honeybee's digestive system is unable to function properly. This can be caused by a number of factors, including:

- ▶ Damage to the gut lining
- ▶ An imbalance of gut bacteria
- ▶ Poor digestion of food

Transmission

Dysentery can be transmitted from bee to bee through contact with contaminated faeces or food. It can also be transmitted through contaminated hive equipment.

Signs of Dysentery

Image courtesy Leila Pulsifer BC Honey Producers Association/Nuria Morfin

Symptoms

The symptoms of dysentery in honeybees include:

- ▶ Diarrhoea
- ▶ Soiling of the hive with faeces
- ▶ Increased mortality of bees
- ▶ Reduced honey production

How to identify dysentery

Dysentery can be identified by the presence of faeces on the hive and on the bees themselves. The faeces will be watery and brown in colour.

Impact on colonies

Dysentery can have a significant impact on honeybee colonies. Heavy infestation can lead to the collapse of a colony. Dysentery can also weaken the colony and make it more susceptible to other problems, such as varroa mite infestation and American foulbrood.

Advice on mitigation

There are a number of things that beekeepers can do to mitigate the impact of dysentery:

- ▶ Keep the hive strong and healthy.
- ▶ Inspect the hive regularly for signs of disease and pests.
- ▶ Provide the bees with a balanced diet.
- ▶ Avoid feeding the bees honey from unknown sources.
- ▶ Harvest honey from the hive as soon as possible after it is capped.
- ▶ Store honey in a cool, dark place.
- ▶ Keep the hive dry and well-ventilated.
- ▶ Avoid overcrowding the hive.
- ▶ Use clean and sanitised hive equipment.

Additional advice on mitigation

Researchers are working to develop new ways to control dysentery in honeybees. One promising approach is to develop probiotics that can help to restore the balance of gut bacteria in bees. Another promising approach is to develop treatments that can target the parasites and diseases that can cause dysentery.

43. Egyptian Acute Paralysis Virus

Egyptian Acute Paralysis Virus (EAPB) was once thought to be a distinct bee virus but recent research has shown it to be a variant of Deformed Wing Virus (DWV) known as DWV-D. DWV is the most common virus affecting honey bees worldwide, and DWV-D was first identified in Egypt in the 1970s.

Classification

- ▶ Class: Monodnaviridae
- ▶ Order: Picornavirales
- ▶ Family: Dicistroviridae
- ▶ Genus: Dicistrovirus [2]

Note: Earlier classifications listed EAPV under the genus Picornavirus.

Biology

DWV-D is a single-stranded RNA virus with an isometric capsid (protein shell) approximately 30nm in diameter [2]. Like other RNA viruses, DWV-D replicates inside the host cell by hijacking its machinery. This replication process can damage the host cell and lead to cell death.

Transmission

DWV-D is primarily transmitted by the Varroa mite (Varroa destructor) which feeds on the hemolymph (insect blood) of honeybees. As the mite feeds, it can pick up the virus from infected bees and then transmit it to healthy bees while feeding. DWV-D can also be transmitted horizontally between bees through contact with contaminated food or faeces.

Symptoms

Honeybees infected with DWV-D may exhibit the following symptoms:

- ▶ Tremors
- ▶ Paralysis
- ▶ Shriveling wings
- ▶ Difficulty flying
- ▶ Disorientation

However, these symptoms can also be caused by other bee viruses, so laboratory testing is necessary to confirm a DWV-D infection.

Identification

DWV-D can be identified using a variety of methods, including:

- ▶ Enzyme-linked immunosorbent assay (ELISA): This is a common laboratory technique that can be used to detect the presence of viral proteins in bee samples.

▶ Polymerase chain reaction (PCR): PCR is a more sensitive technique that can be used to detect the presence of viral RNA in bee samples.

▶ Next-generation sequencing: This technique can be used to identify all of the viruses present in a bee sample, including DWV-D.

Impact on Colonies

DWV-D infection can have a significant impact on honeybee colonies. Infected bees may be unable to forage for food, care for brood, or defend the hive. This can lead to weakened colonies that are more susceptible to other diseases and pests. In severe cases, DWV-D infection can lead to colony collapse disorder (CCD).

Mitigation

There is no cure for DWV-D infection, but there are a number of things that beekeepers can do to help mitigate its effects:

▶ Maintain strong colonies: Strong colonies are better able to tolerate the effects of DWV-D infection. Beekeepers can help to keep their colonies strong by providing them with adequate nutrition and managing Varroa mite populations.

▶ Monitor Varroa mite populations: Regularly monitoring Varroa mite populations is essential for preventing the spread of DWV-D. Beekeepers can use a variety of methods to monitor Varroa mite levels, such as sugar dusting or alcohol washes.

▶ Control Varroa mites: There are a number of Varroa mite control products available, including miticides and organic control methods. Beekeepers should develop an integrated pest management (IPM) plan to control Varroa mites.

▶ Genetic selection: Some bee lines appear to be more resistant to DWV-D infection than others. Beekeepers may want to consider selecting bees from these lines.

By taking these steps, beekeepers can help to reduce the impact of DWV-D on their colonies.

Important Note: While research into bee viruses is ongoing, there is currently no commercially available vaccine or specific treatment for DWV-D. Beekeepers should focus on preventative measures to keep their hives healthy.

44. ESCHERICHA COLI

Escherichia coli (E. coli) is a gram-negative, rod-shaped bacterium that is classified as follows:

Classification

- ▶ Kingdom: Bacteria
- ▶ Phylum: Firmicutes
- ▶ Class: Gammaproteobacteria
- ▶ Order: Enterobacteriales
- ▶ Family: Enterobacteriaceae
- ▶ Genus: Escherichia
- ▶ Species: Escherichia coli

Biology

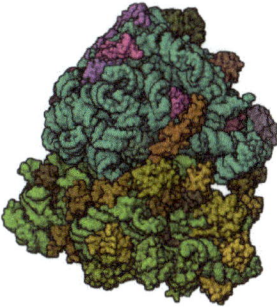

6GW Escherichia Coli

Courtesy rcsb.org

Cryo-EM structure of an E. coli 70S ribosome in complex with RF3-GDPCP, RF1(GAQ) and Pint-tRNA (State I)
PDB DOI: https://doi.org/10.2210/pdb6GWT/pdb
Deposition Author(s): Graf, M., Huter, P., Maracci, C., Peterek, M., Rodnina, M.V., Wilson, D.N.

E. coli is a common bacterium that is found in the environment, including in soil, water and food. It is also found in the intestines of humans and animals, including honeybees.

E. coli is a versatile bacterium that can survive in a variety of conditions. It can grow in both aerobic and anaerobic conditions and it can tolerate a wide range of temperatures and pH levels.

Transmission

E. coli can be transmitted to honeybees through contaminated food, water, or hive equipment. It can also be transmitted by adult bees that carry the bacteria on their bodies.

Once inside the hive, E. coli can infect a bee's gut and nervous system. The bacteria can also cause disease in a bee's brood.

Symptoms

The symptoms of E. coli infection in honeybees include:

- ▶ Diarrhoea
- ▶ Dysentery
- ▶ Death

How to identify E. coli

E. coli can be identified by culturing samples from the hive, such as brood, food and hive equipment, on specialised media. The bacteria can also be identified using molecular methods, such as PCR.

Impact on colonies

E. coli infection can have a significant impact on honeybee colonies. Heavy infestation can lead to the collapse of a colony. E. coli infection can also weaken the colony and make it more susceptible to other problems, such as varroa mite infestation and American foulbrood.

Advice on mitigation

There is no cure for E. coli infection in honeybees. However, there are a number of things that beekeepers can do to mitigate the impact of the disease:

▸ Keep the hive strong and healthy.
▸ Inspect the hive regularly for signs of disease.
▸ Avoid overcrowding the hive.
▸ Use clean and sanitised hive equipment.
▸ Avoid introducing new bees into the hive without first inspecting them for disease.
▸ Purchase bees from reputable beekeepers.
▸ Avoid feeding the bees honey from unknown sources.
▸ Harvest honey from the hive as soon as possible after it is capped.
▸ Store honey in a cool, dark place.

Additional advice on mitigation

Researchers are working to develop new ways to control E. coli infection in honeybees. One promising approach is to develop probiotics that can compete with E. coli in the bees' gut. Another promising approach is to develop vaccines that can protect the bees from E. coli infection.

45. ENTEROCOCCUS FAECALIS

Enterococcus faecalis is a gram-positive, cocci-shaped bacterium that is classified as follows:

Classification

- Kingdom: Bacteria
- Phylum: Firmicutes
- Class: Bacilli
- Order: Lactobacillales
- Family: Enterococcaceae
- Genus: Enterococcus
- Species: Enterococcus faecalis

Biology

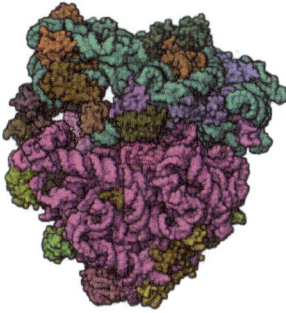

6O8X Enterococcus Faecalis Cryo-EM image reconstruction of the 70S Ribosome Enterococcus faecalis Class02

PDB DOI: https://doi.org/10.2210/pdb6O8X/pdb

Deposition Author(s): Jogl, G., Khayat, R.

Courtesy rcsb.org

E. faecalis is a common bacterium that is found in the environment, including in soil, water and food. It is also found in the intestines of humans and animals, including honeybees.

E. faecalis is a versatile bacterium that can survive in a variety of conditions. It can grow in both aerobic and anaerobic conditions and it can tolerate a wide range of temperatures and pH levels.

Transmission

E. faecalis can be transmitted to honeybees through contaminated food, water, or hive equipment. It can also be transmitted by adult bees that carry the bacteria on their bodies.

Once inside the hive, E. faecalis can colonise the bees' gut and other tissues. The bacteria can also cause disease in the bees' brood.

Symptoms

The symptoms of E. faecalis infection in honeybees include:

- Diarrhoea
- Dysentery
- Death
- Stunted growth in brood

How to identify E. faecalis

E. faecalis can be identified by culturing samples from the hive, such as brood, food and hive equipment, on specialised media. The bacteria can also be identified using molecular methods, such as PCR.

Impact on colonies

E. faecalis infection can have a significant impact on honeybee colonies. Heavy infestation can lead to the collapse of a colony. E. faecalis infection can also weaken the colony and make it more susceptible to other problems, such as varroa mite infestation and American foulbrood.

Advice on mitigation

There is no cure for E. faecalis infection in honeybees. However, there are a number of things that beekeepers can do to mitigate the impact of the disease:

▸ Keep the hive strong and healthy.
▸ Inspect the hive regularly for signs of disease.
▸ Avoid overcrowding the hive.
▸ Use clean and sanitised hive equipment.
▸ Avoid introducing new bees into the hive without first inspecting them for disease.
▸ Purchase bees from reputable beekeepers.
▸ Avoid feeding the bees honey from unknown sources.
▸ Harvest honey from the hive as soon as possible after it is capped.
▸ Store honey in a cool, dark place.

Additional advice on mitigation

Researchers are working to develop new ways to control E. faecalis infection in honeybees. One promising approach is to develop probiotics that can compete with E. faecalis in the bees' gut. Another promising approach is to develop vaccines that can protect the bees from E. faecalis infection.

46. EUROPEAN FOUL BROOD

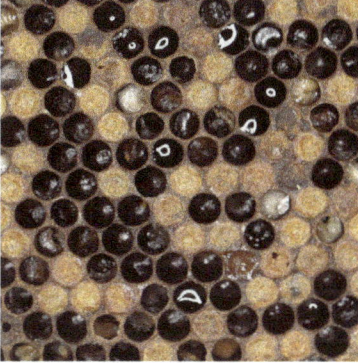

European Foulbrood (EFB)
Advance stage of EFB infection in comb with "melted", discoloured, slumped larvae.

Courtesy the Animal and Plant Health Agency © Crown Copyright

European Foulbrood (EFB) melted down larvae

Courtesy the Animal and Plant Health Agency © Crown Copyright

See melissococcus plutonius

47. FILAMENTOUS VIRUS

The filamentous virus in honeybees is known as Apis mellifera filamentous virus (AmFV). It is a large, double-stranded DNA virus that is classified as follows:

Classification

- Kingdom: Viruses
- Phylum: Riboviria
- Class: Picornavirales
- Order: Picornavirales
- Family: Iflaviridae
- Genus: Iflavirus
- Species: Apis mellifera filamentous virus

Biology

AmFV is a relatively large virus, with a genome that is about 3000 nanometers long. The virus replicates in the cytoplasm of honeybee cells. The first step in the replication cycle is the attachment of the virus to the cell surface. The virus then enters the cell and its DNA is transcribed into RNA. The RNA is then translated into proteins, which form the new virus particles.

Transmission

AmFV is transmitted from bee to bee through contact with infected bees or their faeces. The virus can also be transmitted through contaminated hive equipment.

Symptoms

AmFV infection is generally asymptomatic in honeybees. However, heavy infestation can lead to the following symptoms:

- Reduced honey production
- Increased mortality of bees
- Weakened colony

How to identify AmFV

AmFV infection can be identified by testing honeybee samples for the presence of the virus's DNA. This can be done using a variety of methods, such as real-time PCR (qPCR) or reverse transcription PCR (RT-PCR).

Impact on colonies

AmFV infection can have a significant impact on honeybee colonies. Heavy infestation can lead to the collapse of a colony. AmFV infection can also weaken the colony and make it more susceptible to other problems, such as varroa mite infestation and American foulbrood.

Advice on mitigation

There is no cure for AmFV infection in honeybees. However, there are a number of things that beekeepers can do to mitigate the impact of the disease:

- ▶ Keep the hive strong and healthy.
- ▶ Inspect the hive regularly for signs of disease.
- ▶ Avoid overcrowding the hive.
- ▶ Use clean and sanitised hive equipment.
- ▶ Avoid introducing new bees into the hive without first inspecting them for disease.
- ▶ Purchase bees from reputable beekeepers.
- ▶ Requeen the colony if it is heavily infested with AmFV.

Additional advice on mitigation

Researchers are working to develop new ways to control AmFV infection. One promising approach is to develop vaccines against the virus. Another promising approach is to use RNA interference (RNAi) to silence the genes that are responsible for the virus's virulence.

48. FLAVIVIRUS

Flavirus is an enveloped positive-stranded dNA virus which is classified as follows

Classification:

▸ Kingdom: Viruses
▸ Realm: Riboviridae
▸ Phylum: Negarnaviricota
▸ Class: Amarillovirales
▸ Order: Flaviviridae
▸ Genus: Flavivirus

Biology:

Flaviviruses are enveloped, positive-stranded RNA viruses that primarily infect mammals and birds. They are transmitted through arthropod vectors, such as ticks and mosquitoes. The family name "Flaviviridae" comes from the yellow fever virus; "flavus" is Latin for "yellow," and yellow fever is named due to its propensity to cause jaundice in victims. There are 89 recognised species in the family, divided among four genera. Flaviviridae includes viruses that cause significant diseases in humans, including hepatitis C, dengue fever, Japanese encephalitis, West Nile virus and Sika virus.

Transmission

Flaviviruses can infect honeybees through contact with contaminated food, water, or hive equipment. They can also be transmitted by Varroa mites, which are parasitic mites that feed on honeybees.

Symptoms:

The symptoms of Flavivirus infection in honeybees can vary depending on the strain of virus involved and the severity of the infection. Some of the most common symptoms include:

▸ Reduced honey production
▸ Increased mortality
▸ Deformed wings
▸ Paralysis
▸ Hair loss
▸ Darkened abdomens
▸ Tremors

Identification:

Flaviviruses can be identified by laboratory testing of honeybee samples. The most common methods of detection include:

▸ Reverse transcription-polymerase chain reaction (RT-PCR): This test can detect the presence of Flavivirus RNA in honeybee samples.

- Ensyme-linked immunosorbent assay (ELISA): This test can detect the presence of Flavivirus antibodies in honeybee samples.

Impact on Colonies:

Flavivirus infection can have a significant impact on honeybee colonies. Heavy infection can lead to:
- Reduced honey production
- Increased mortality
- Colony collapse

Mitigation:

There is no cure for Flavivirus infection in honeybees. However, there are a number of things that beekeepers can do to mitigate the impact of the viruses:
- Maintain strong colony health: Strong colonies are better able to withstand the effects of Flavivirus infection.
- Provide adequate nutrition: Ensure that the colony has access to a plentiful supply of pollen and other essential nutrients.
- Control hive temperature: Avoid extreme temperatures in the hive by providing adequate ventilation and insulation.
- Avoid pesticide exposure: Use pesticides only as a last resort and follow label directions carefully.
- Monitor colony health: Regularly inspect the colony for signs of Flavivirus infection, such as reduced honey production, increased mortality, deformed wings, paralysis, hair loss, darkened abdomens and tremors.
- Isolate infected colonies: If you find Flavivirus infection in a colony, isolate it from other colonies to prevent the spread of the viruses.
- Destroy infected hives: If a colony is heavily infected with Flavivirus, it is best to destroy the hive and all its contents.
- Purchase bees from reputable beekeepers: Purchase bees from reputable beekeepers who have taken steps to control Flavivirus in their colonies.

49. FUSARIUM OXYSPORUM

Similar to the broader genus Fusarium, Fusarium oxysporum is not a direct pathogen of honeybees (Apis mellifera). This specific fungal species primarily targets vascular tissues in plants, causing wilt diseases. However, its presence can indirectly impact bee health through contaminated food sources and a compromised hive environment.

Classification

- ▶ Kingdom: Fungi
- ▶ Phylum: Ascomycota
- ▶ Class: Sordariomycetes
- ▶ Order: Hypocreales
- ▶ Genus: Fusarium
- ▶ Species: Fusarium oxysporum

Biology

- ▶ Fusarium oxysporum is a soilborne fungus with a wide host range, infecting various plants.
- ▶ It colonises plant roots, secretes toxins, and disrupts the vascular system, leading to wilting and plant death.
- ▶ The fungus reproduces by forming spores that can survive in the soil for extended periods.

Impact on Honeybees

The primary concern for honeybees is not direct infection but the indirect effects of F. oxysporum on:

- ▶ Contaminated Pollen:
  - • F. oxysporum can infect flowers and contaminate pollen, a vital food source for bees.
  - • Contaminated pollen can be:
    - ▪ Toxic to developing bee larvae.
    - ▪ Less nutritious, hindering larval growth and development.
  - • Spores on pollen can be carried back to the hive, potentially contributing to fungal growth within.
- ▶ Weakened Hive Environment:
  - • F. oxysporum growth within the hive on:
    - ▪ Dead bees
    - ▪ Brood comb debris
    - ▪ Stored food (if it's plant-based)
  - • This can:
    - ▪ Increase humidity levels, promoting the growth of other harmful molds

and fungi.

- Create an unhealthy and unsanitary environment for the bees.

Symptoms

Honeybees themselves won't exhibit specific symptoms directly related to F. oxysporum. However, the indirect effects can manifest as:

▶ Reduced brood production: Poor quality pollen or a compromised hive environment can lead to lower brood rearing activity.

▶ Increased adult bee mortality: Weakened bees due to nutritional deficiencies or a compromised hive environment may be more susceptible to diseases and other stressors.

▶ •Colony decline: A combination of these factors can lead to a gradual decline in overall colony health and population.

Identification

Identifying F. oxysporum specifically in the context of honeybees is challenging. Visual inspection may reveal signs of fungal growth on pollen or within the hive, but laboratory analysis is necessary for confirmation.

Mitigation Strategies

▶ Promote strong, healthy colonies: Strong colonies are better equipped to handle challenges like pollen contamination.

▶ Good hive management practices:
  - Regular inspections
  - Removal of debris
  - Proper ventilation
  - These practices help prevent moisture buildup and create an unfavourable environment for fungal growth.

▶ Provide high-quality pollen sources: Planting diverse bee-friendly flowers or providing supplemental pollen patties helps ensure access to good quality food.

▶ Monitor for other stressors: Addressing underlying issues like Varroa mites or other diseases that weaken bees can improve their resilience to F. oxysporum's indirect effects.

Important Note:

While F. oxysporum presents an indirect threat, it's unlikely to be the primary cause of honeybee colony decline. A comprehensive beekeeping approach that addresses various stressors and promotes optimal hive health is essential for mitigating the overall impact of F. oxysporum and other environmental factors.

50. INVERTEBRATE IRIDESCENT VIRUS TYPE 6

Invertebrate iridescent virus type 6 is a large enveloped DNA virus which is classified as follows

Classification

- ▶ Kingdom: Riboviridae
- ▶ Phylum: Negarnaviricota
- ▶ Class: Picornaviridae
- ▶ Order: Picornavirales
- ▶ Family: Iflaviridae
- ▶ Genus: Iflavirus
- ▶ Species: Sacbrood virus

Biology

IIV-6 is a large, enveloped DNA virus with a capsid that measures approximately 120-130 nm in diameter. The virus replicates in the cytoplasm of honeybee cells, producing iridescent inclusions that give the virus its name. IIV-6 can infect a wide range of honeybee tissues, including the epidermis, midgut, fat body and flight muscles.

Transmission

IIV-6 is primarily transmitted from bee to bee through contact with contaminated faeces or saliva. Infected adult bees can shed the virus through their faeces or saliva, which can contaminate food and water sources for other bees. IIV-6 can also be transmitted by Varroa mites, which are parasitic mites that feed on honeybees.

Symptoms

The symptoms of IIV-6 infection can vary depending on the severity of the infection and the age of the bee. Some of the most common symptoms include:

- ▶ Reduced honey production
- ▶ Increased mortality
- ▶ Deformed wings
- ▶ Paralysis
- ▶ Hair loss
- ▶ Darkened abdomens
- ▶ Sticky larvae
- ▶ Swollen abdomens
- ▶ Discoloured pupae
- ▶ Stunted growth
- ▶ Poor flying ability

Identification

IIV-6 can be identified by laboratory testing of honeybee samples. The most common methods of detection include:

- Electron microscopy: This method can visualise the virus particles in honeybee tissues.
- Polymerase chain reaction (PCR): This test can detect the presence of IIV-6 DNA in honeybee samples.
- Ensyme-linked immunosorbent assay (ELISA): This test can detect the presence of IIV-6 antibodies in honeybee samples.

Impact on Colonies

IIV-6 can have a significant impact on honeybee colonies. Heavy infection can lead to:

- Reduced honey production
- Increased mortality
- Colony collapse

Advice on Mitigation

There is no cure for IIV-6 infection in honeybees. However, there are a number of things that beekeepers can do to mitigate the impact of the virus. These include:

- Maintain strong colony health: Strong colonies are better able to withstand the effects of IIV-6.
- Provide adequate nutrition: Ensure that the colony has access to a plentiful supply of pollen and other essential nutrients.
- Control hive temperature: Avoid extreme temperatures in the hive by providing adequate ventilation and insulation.
- Avoid pesticide exposure: Use pesticides only as a last resort and follow label directions carefully.
- Monitor colony health: Regularly inspect the colony for signs of IIV-6 infection, such as reduced honey production, increased mortality, deformed wings, paralysis, hair loss, darkened abdomens, sticky larvae, swollen abdomens and discoloured pupae.
- Isolate infected colonies: If you find IIV-6 infection in a colony, isolate it from other colonies to prevent the spread of the virus.
- Destroy infected hives: If a colony is heavily infected with IIV-6, it is best to destroy the hive and all its contents.
- Purchase bees from reputable beekeepers: Purchase bees from reputable beekeepers who have taken steps to control IIV-6 infection in their colonies.
- Monitor Varroa mite infestation: Control Varroa mite infestation as Varroa mites can transmit the virus.

51. ISRAELI ACUTE PARALYSIS VIRUS

Israeli acute paralysis virus is a small RNA virus which is classified as follows

Classification

- ▶ Kingdom: Riboviridae
- ▶ Phylum: Negarnaviricota
- ▶ Class: Picornaviridae
- ▶ Order: Picornavirales
- ▶ Family: Dicistroviridae
- ▶ Genus: Dicistrovirus
- ▶ Species: Israeli acute paralysis virus

Biology

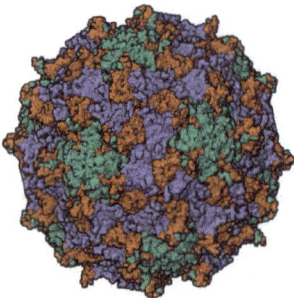

Courtesy rcsb.org

5CDC Crystal Structure of Israel acute Paralysis Virus

PDB DOI: https://doi.org/10.2210/pdb5CDC/pdb

Deposition Author(s): Mullapudi, E., Plevka, P.

IAPV is a small, RNA virus with a genome of approximately 2,500 nucleotides. The virus replicates in the cytoplasm of honeybee cells, producing two proteins that are essential for its replication. IAPV can infect a wide range of honeybee tissues, including the midgut, nervous system and hypopharyngeal glands.

Transmission

IAPV is transmitted from bee to bee through contact with contaminated faeces or saliva. Infected adult bees can shed the virus through their faeces or saliva, which can contaminate food and water sources for other bees. IAPV can also be transmitted by Varroa mites, which are parasitic mites that feed on honeybees.

Symptoms

The symptoms of IAPV infection can vary depending on the severity of the infection and the age of the bee. Some of the most common symptoms include:

- ▶ Shivering wings
- ▶ Cramping
- ▶ Disorientation
- ▶ Paralysis
- ▶ Death

Paralysis typically occurs within a few days of infection.

Identification

IAPV can be identified by laboratory testing of honeybee samples. The most common methods of detection include:

- Polymerase chain reaction (PCR): This test can detect the presence of IAPV RNA in honeybee samples.
- Ensyme-linked immunosorbent assay (ELISA): This test can detect the presence of IAPV antibodies in honeybee samples.
- Electron microscopy: This method can visualise the virus particles in honeybee tissues.

Impact on Colonies

IAPV can have a significant impact on honeybee colonies. Heavy infection can lead to:

- Reduced honey production
- Increased mortality
- Colony collapse

Advice on Mitigation

There is no cure for IAPV infection in honeybees. However, there are a number of things that beekeepers can do to mitigate the impact of the virus. These include:

- Maintain strong colony health: Strong colonies are better able to withstand the effects of IAPV.
- Provide adequate nutrition: Ensure that the colony has access to a plentiful supply of pollen and other essential nutrients.
- Control hive temperature: Avoid extreme temperatures in the hive by providing adequate ventilation and insulation.
- Avoid pesticide exposure: Use pesticides only as a last resort and follow label directions carefully.
- Monitor colony health: Regularly inspect the colony for signs of IAPV infection, such as shivering wings, cramping, disorientation, paralysis and death.
- Isolate infected colonies: If you find IAPV infection in a colony, isolate it from other colonies to prevent the spread of the virus.
- Destroy infected hives: If a colony is heavily infected with IAPV, it is best to destroy the hive and all its contents.
- Purchase bees from reputable beekeepers: Purchase bees from reputable beekeepers who have taken steps to control IAPV infection in their colonies.

52. ISRAELI ACUTE PARALYSIS-LIKE VIRUS

Israeli Acute Paralysis Virus-like Virus (IAPVLV) is a viral pathogen of honeybees that shares many similarities with Israeli Acute Paralysis Virus (IAPV). However, IAPVLV is genetically distinct from IAPV and is considered to be a separate virus.

Classification

Kingdom: Riboviridae
- ▸ Phylum: Negarnaviricota
- ▸ Class: Picornaviridae
- ▸ Order: Picornavirales
- ▸ Family: Dicistroviridae
- ▸ Genus: Dicistrovirus
- ▸ Species: Israeli acute paralysis virus-like virus (IAPVLV)

Biology

IAPVLV is a small, RNA virus with a genome of approximately 2,500 nucleotides. The virus replicates in the cytoplasm of honeybee cells, producing two proteins that are essential for its replication. IAPVLV can infect a wide range of honeybee tissues, including the midgut, nervous system and hypopharyngeal glands.

Transmission

IAPVLV is transmitted from bee to bee through contact with contaminated faeces or saliva. Infected adult bees can shed the virus through their faeces or saliva, which can contaminate food and water sources for other bees. IAPVLV can also be transmitted by Varroa mites, which are parasitic mites that feed on honeybees.

Symptoms

The symptoms of IAPVLV infection are similar to those of IAPV and can vary depending on the severity of the infection and the age of the bee. Some of the most common symptoms include:
- ▸ Shivering wings
- ▸ Cramping
- ▸ Disorientation
- ▸ Paralysis
- ▸ Death

Paralysis typically occurs within a few days of infection.

Identification

IAPVLV can be identified by laboratory testing of honeybee samples. The most common methods of detection include:

- Polymerase chain reaction (PCR): This test can detect the presence of IAPVLV RNA in honeybee samples.
- Ensyme-linked immunosorbent assay (ELISA): This test can detect the presence of IAPVLV antibodies in honeybee samples.
- Electron microscopy: This method can visualise the virus particles in honeybee tissues.

Impact on Colonies

IAPVLV can have a significant impact on honeybee colonies. Heavy infection can lead to:

- Reduced honey production
- Increased mortality
- Colony collapse

Advice on Mitigation

There is no cure for IAPVLV infection in honeybees. However, there are a number of things that beekeepers can do to mitigate the impact of the virus. These include:

- Maintain strong colony health: Strong colonies are better able to withstand the effects of IAPVLV.
- Provide adequate nutrition: Ensure that the colony has access to a plentiful supply of pollen and other essential nutrients.
- Control hive temperature: Avoid extreme temperatures in the hive by providing adequate ventilation and insulation.
- Avoid pesticide exposure: Use pesticides only as a last resort and follow label directions carefully.
- Monitor colony health: Regularly inspect the colony for signs of IAPVLV infection, such as shivering wings, cramping, disorientation, paralysis and death.
- Isolate infected colonies: If you find IAPVLV infection in a colony, isolate it from other colonies to prevent the spread of the virus.
- Destroy infected hives: If a colony is heavily infected with IAPVLV, it is best to destroy the hive and all its contents.
- Purchase bees from reputable beekeepers: Purchase bees from reputable beekeepers who have taken steps to control IAPVLV infection in their colonies.
- Monitor Varroa mite infestation: Control Varroa mite infestation as Varroa mites can transmit the virus.

53. KAKUGO VIRUS

Kakugo virus (KV) is a single-stranded RNA virus that is classified as follows:

Classification

- ▶ Kingdom: Viruses
- ▶ Phylum: Riboviria
- ▶ Class: Picornavirales
- ▶ Order: Iflaviridae
- ▶ Family: Iflaviridae
- ▶ Genus: Iflavirus
- ▶ Species: Kakugo virus

Biology

KV is a relatively small virus, with a genome that is about 8,000 nucleotides long. The virus replicates in the cytoplasm of honeybee cells. The first step in the replication cycle is the attachment of the virus to the cell surface. The virus then enters the cell and its RNA is translated into proteins. The proteins form a complex that replicates the viral RNA. The new viral RNA is then translated into proteins, which form the new virus particles.

Transmission

KV is transmitted from bee to bee through contact with infected bees or their faeces. The virus can also be transmitted through contaminated hive equipment.

Symptoms

The symptoms of KV infection in honeybees include:

- ▶ Aggression
- ▶ Increased mortality
- ▶ Reduced honey production

How to identify KV

KV infection can be identified by testing honeybee samples for the presence of the virus's RNA. This can be done using a variety of methods, such as real-time PCR (qPCR) or reverse transcription PCR (RT-PCR).

Impact on colonies

KV infection can have a significant impact on honeybee colonies. Heavy infestation can lead to the collapse of a colony. KV infection can also weaken the colony and make it more susceptible to other problems, such as varroa mite infestation and American foulbrood.

Advice on mitigation

There is no cure for KV infection in honeybees. However, there are a number of things that beekeepers can do to mitigate the impact of the disease:

▶ Keep the hive strong and healthy.
▶ Inspect the hive regularly for signs of disease.
▶ Avoid overcrowding the hive.
▶ Use clean and sanitised hive equipment.
▶ Avoid introducing new bees into the hive without first inspecting them for disease.
▶ Purchase bees from reputable beekeepers.
▶ Requeen the colony if it is heavily infested with KV.

Additional advice on mitigation

Researchers are working to develop new ways to control KV infection. One promising approach is to develop vaccines against the virus. Another promising approach is to use RNA interference (RNAi) to silence the genes that are responsible for the virus's virulence.

54. KASHMIR BEE VIRUS

Kashmir bee virus (KBV) is a single-stranded RNA virus that is classified as follows

Classification:

▸ Kingdom: Viruses
▸ Phylum: Riboviria
▸ Class: Picornavirales
▸ Order: Iflaviridae
▸ Family: Iflaviridae
▸ Genus: Cripavirus
▸ Species: Kashmir bee virus

Biology

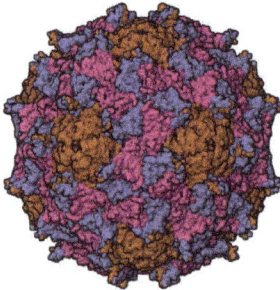

7BC3 Native virion of Kashmir bee virus at acidic Ph

Courtesy rcsb.org
PDB DOI: https://doi.org/10.2210/pdb7BC3/pdb
Deposition Author(s): Mukhamedova, L., Plevka, P., Fuzik, T., Hrebik, D., Novacek, J.

KBV is a relatively small virus, with a genome that is about 8,500 nucleotides long. The virus replicates in the cytoplasm of honeybee cells. The first step in the replication cycle is the attachment of the virus to the cell surface. The virus then enters the cell and its RNA is translated into proteins. The proteins form a complex that replicates the viral RNA. The new viral RNA is then translated into proteins, which form the new virus particles.

Transmission

KBV is transmitted from bee to bee through contact with infected bees or their faeces. The virus can also be transmitted through contaminated hive equipment.

Symptoms

The symptoms of KBV infection in honeybees include:

▸ Paralysis
▸ Weakness
▸ Lethargy
▸ Reduced honey production
▸ Death

How to identify KBV

KBV infection can be identified by testing honeybee samples for the presence of the virus's RNA. This can be done using a variety of methods, such as real-time PCR (qPCR) or reverse transcription PCR (RT-PCR).

Impact on colonies

KBV infection can have a significant impact on honeybee colonies. Heavy infestation can lead to the collapse of a colony. KBV infection can also weaken the colony and make it more susceptible to other problems, such as varroa mite infestation and American foulbrood.

Advice on mitigation

There is no cure for KBV infection in honeybees. However, there are a number of things that beekeepers can do to mitigate the impact of the disease:

▶ Keep the hive strong and healthy.
▶ Inspect the hive regularly for signs of disease.
▶ Avoid overcrowding the hive.
▶ Use clean and sanitised hive equipment.
▶ Avoid introducing new bees into the hive without first inspecting them for disease.
▶ Purchase bees from reputable beekeepers.
▶ Requeen the colony if it is heavily infested with KBV.

Additional advice on mitigation

Researchers are working to develop new ways to control KBV infection. One promising approach is to develop vaccines against the virus. Another promising approach is to use RNA interference (RNAi) to silence the genes that are responsible for the virus's virulence.

55. KASHMIR BEE VIRUS-LIKE VIRUS

Kashmir bee virus-like virus (KBV-LV) is a single-stranded RNA virus that is closely related to Kashmir bee virus (KBV). However, KBV-LV is a distinct virus and is classified as follows:

Classification

▶ Kingdom: Viruses
▶ Phylum: Riboviria
▶ Class: Picornavirales
▶ Order: Iflaviridae
▶ Family: Iflaviridae
▶ Genus: Cripavirus
▶ Species: Kashmir bee virus-like virus

Biology

KBV-LV is a relatively small virus, with a genome that is about 8,500 nucleotides long. The virus replicates in the cytoplasm of honeybee cells. The first step in the replication cycle is the attachment of the virus to the cell surface. The virus then enters the cell and its RNA is translated into proteins. The proteins form a complex that replicates the viral RNA. The new viral RNA is then translated into proteins, which form the new virus particles.

Transmission

KBV-LV is transmitted from bee to bee through contact with infected bees or their faeces. The virus can also be transmitted through contaminated hive equipment.

Symptoms

The symptoms of KBV-LV infection in honeybees are similar to those of KBV infection and include:

▶ Paralysis
▶ Weakness
▶ Lethargy
▶ Reduced honey production
▶ Death

How to identify KBV-LV

KBV-LV infection can be identified by testing honeybee samples for the presence of the virus's RNA. This can be done using a variety of methods, such as real-time PCR (qPCR) or reverse transcription PCR (RT-PCR).

Impact on colonies

KBV-LV infection can have a significant impact on honeybee colonies. Heavy infestation can lead to the collapse of a colony. KBV-LV infection can also weaken the colony and make it more susceptible to other problems, such as varroa mite infestation and American foulbrood.

Advice on mitigation

There is no cure for KBV-LV infection in honeybees. However, there are a number of things that beekeepers can do to mitigate the impact of the disease:

- Keep the hive strong and healthy.
- Inspect the hive regularly for signs of disease.
- Avoid overcrowding the hive.
- Use clean and sanitised hive equipment.
- Avoid introducing new bees into the hive without first inspecting them for disease.
- Purchase bees from reputable beekeepers.
- Requeen the colony if it is heavily infested with KBV-LV.

Additional advice on mitigation

Researchers are working to develop new ways to control KBV-LV infection. One promising approach is to develop vaccines against the virus. Another promising approach is to use RNA interference (RNAi) to silence the genes that are responsible for the virus's virulence.

56. LAKE SINAI VIRUS 1

Lake Sinai virus (LSV11) is a single-stranded RNA virus that is classified as follows:

Classification

- ▸ Kingdom: Viruses
- ▸ Phylum: Riboviria
- ▸ Class: Picornavirales
- ▸ Order: Iflaviridae
- ▸ Family: Iflaviridae
- ▸ Genus: Sinaivirus
- ▸ Species: Lake Sinai virus

Biology

LSV1 is a relatively small virus, with a genome that is about 5,500 nucleotides long. The virus replicates in the cytoplasm of honeybee cells. The first step in the replication cycle is the attachment of the virus to the cell surface. The virus then enters the cell and its RNA is translated into proteins. The proteins form a complex that replicates the viral RNA. The new viral RNA is then translated into proteins, which form the new virus particles.

Transmission

LSV1 is transmitted from bee to bee through contact with infected bees or their faeces. The virus can also be transmitted through contaminated hive equipment.

Symptoms

The symptoms of LSV1 infection in honeybees are not well-defined. However, some studies have shown that LSV1 infection can be associated with:

- ▸ Reduced honey production
- ▸ Increased mortality of bees
- ▸ Weakened colony

How to identify LSV1

LSV1 infection can be identified by testing honeybee samples for the presence of the virus's RNA. This can be done using a variety of methods, such as real-time PCR (qPCR) or reverse transcription PCR (RT-PCR).

Impact on colonies

The impact of LSV1 infection on honeybee colonies is not fully understood. However, some studies have shown that LSV1 infection can be associated with:

- ▸ Reduced honey production
- ▸ Increased mortality of bees
- ▸ Weakened colony

▶ Collapse of the colony

Advice on mitigation

There is no cure for LSV1 infection in honeybees. However, there are a number of things that beekeepers can do to mitigate the impact of the disease:

▶ Keep the hive strong and healthy.
▶ Inspect the hive regularly for signs of disease.
▶ Avoid overcrowding the hive.
▶ Use clean and sanitised hive equipment.
▶ Avoid introducing new bees into the hive without first inspecting them for disease.
▶ Purchase bees from reputable beekeepers.
▶ Requeen the colony if it is heavily infested with LSV1.

Additional advice on mitigation

Researchers are working to develop new ways to control LSV1 infection. One promising approach is to develop vaccines against the virus. Another promising approach is to use RNA interference (RNAi) to silence the genes that are responsible for the virus's virulence.

57. LAKE SINAI VIRUS 2

Lake Sinai virus 2 (LSV2) is one of several members of the Lake Sinai virus (LSV1) family, a diverse group of single-stranded RNA viruses infecting honeybees (Apis mellifera). While not causing distinct, easily identifiable symptoms, LSV2 can contribute to colony weakening and deserves attention.

Classification:

▶ Domain: Riboviria
▶ Phylum: Positiscripta
▶ Class: Picornaviricetes
▶ Order: Nodamuvirales
▶ Family: Siphoviridae
▶ Subfamily: Alphabaculovirinae
▶ Genus: Sinaivirus
▶ Species: Lake Sinai virus

Biology

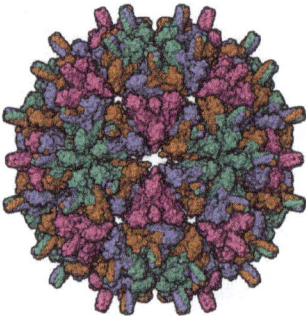

Lake Sinai Virus-2

Courtesy rcsb.org

7XGZ Cryo-EM structure of the T=4 lake sinai virus 2 virus-like capsid at pH 7.5

PDB DOI: https://doi.org/10.2210/pdb7XGZ/pdb

Deposition Author(s): Chen, N.C., Wang, C.H., Chen, C.J., Yoshimura, M., Guan, H.H., Chuankhayan, P., Lin, C.C.

LSV2 has a segmented, single-stranded RNA genome and exists in multiple genetic variants. It replicates within bee cells, primarily targeting the digestive system and potentially other tissues.

Transmission:

Horizontal transmission:

Worker-to-worker contact: During grooming, feeding, or other interactions.

Contaminated hive equipment: Frames, tools, and feeders can harbor and spread the virus.

Vertical transmission:

Infected queens can potentially transmit the virus to their offspring through infected eggs, although this is less common than horisontal transmission.

Symptoms:

Limited specific symptoms: LSV2 often doesn't cause distinct, easily identifiable symptoms in infected bees.

Potential contribution to colony weakening: Studies suggest LSV2 may contribute to reduced bee lifespan, foraging efficiency, and overall colony strength.

Increased susceptibility to other diseases: Weakened colonies due to LSV2 infection might be more vulnerable to other pathogens and parasites.

Identification:

Molecular tests: Polymerase Chain Reaction (PCR) or other techniques can detect the virus's genetic material in honey, bees, or hive debris.

Virus isolation and characterisation: Advanced laboratory techniques can isolate and characterise the virus from infected bees.

Colony observations: While not diagnostic, weakened colonies with decreased foraging activity and increased mortality might raise suspicion of LSV2 infection, requiring further investigation.

Impact on Colonies:

Potential reduction in colony strength: LSV2 infection may contribute to a weakened workforce, reduced honey production, and increased susceptibility to other diseases.

Sublethal effects: Unlike some bee viruses causing rapid mortality, LSV2 often exhibits sublethal effects, impacting long-term colony health and productivity.

Uncertain long-term consequences: The full impact of LSV2 on honeybee health and colony survival is still being investigated.

Mitigation Strategies:

Maintaining strong, hygienic colonies is crucial for overall bee health and resistance to various pathogens.

Good beekeeping practices like regular hive inspections, equipment hygiene, and appropriate nutrition can contribute to colony resilience.

Monitoring: Regular monitoring for LSV2 presence through PCR testing can help track its prevalence and inform management decisions.

Limited treatment options: No specific antiviral treatments are available for honeybees. Supportive care like feeding and hive hygiene can help maintain colony health.

Research and development: Ongoing research on LSV2 biology, transmission, and potential mitigation strategies is crucial for better management of this virus in the future. LSV2 is still an actively researched virus, and the understanding of its impact and management continues to evolve. Consulting with a qualified beekeeping expert for specific advice tailored to your situation and local regulations is always recommended.

58. LEPTOSPIROSIS

See Weil's disease

59. MELISSOCOCCUS ALVEI

Melissococcus alvei is frequently found in honeybee colonies that have been weakened by EFB. It acts as a secondary invader, meaning it takes advantage of the already compromised larvae but doesn't initiate the disease itself. The presence of M. alvei in EFB-affected colonies can sometimes lead to misdiagnosis or confusion about the exact cause of the disease.

60. MELISSOCOCCUS PLUTONIUS

Melissococcus plutonius is a Gram-positive bacterium that is classified as follows:

Classification

- Kingdom: Bacteria
- Phylum: Firmicutes
- Class: Bacilli
- Order: Lactobacillales
- Family: Enterococcaceae
- Genus: Melissococcus
- Species: Melissococcus plutonius

Biology

Melissococcus plutonius is a small, spherical bacterium that is about 0.5-1 micrometer in diameter. The bacteria are non-motile and non-sporulating. They are also facultative anaerobes, which means that they can grow in both the presence and absence of oxygen.

Melissococcus plutonius is the causative agent of European foulbrood (EFB), a disease of honeybee larvae. EFB is one of the most common and destructive diseases of honeybees.

Transmission

Melissococcus plutonius is transmitted from bee to bee through contact with infected bees or their faeces. The bacteria can also be transmitted through contaminated hive equipment.

Symptoms

The symptoms of EFB include:
- Deformed larvae
- Discoloured larvae
- Stunted larvae
- Dead larvae
- A sour odour from the hive

How to identify Melissococcus plutonius

Melissococcus plutonius can be identified by testing honeybee samples for the presence of the bacteria. This can be done using a variety of methods, such as culturing the bacteria on a petri dish or using a PCR test.

Impact on colonies

European foulbrood can have a significant impact on honeybee colonies. Heavy infestation can lead to the collapse of the colony. European foulbrood can also weaken the colony and make it more susceptible to other problems, such as varroa mite infestation and American foulbrood.

Advice on mitigation

There is no cure for European foulbrood. However, there are a number of things that beekeepers can do to mitigate the impact of the disease:

- ▶ Keep the hive strong and healthy.
- ▶ Inspect the hive regularly for signs of disease.
- ▶ Avoid overcrowding the hive.
- ▶ Use clean and sanitised hive equipment.
- ▶ Avoid introducing new bees into the hive without first inspecting them for disease.
- ▶ Purchase bees from reputable beekeepers.
- ▶ Requeen the colony if it is heavily infested with Melissococcus plutonius.

Additional advice on mitigation

Researchers are working to develop new ways to control European foulbrood. One promising approach is to develop vaccines against Melissococcus plutonius. Another promising approach is to use bacteriophages, which are viruses that specifically infect bacteria, to kill Melissococcus plutonius.

61. MELITTIN VIRUS

Melittin virus is a single-stranded RNA virus which is classified as follows

Classification

- Kingdom: Viruses
- Realm: Riboviria
- Phylum: Negarnaviricota
- Class: Picornaviricetes
- Order: Picornavirales
- Family: Iflaviridae
- Genus: Iflavirus

Biology

Melittin virus (MLV) is a single-stranded RNA virus that belongs to the family Iflaviridae. It is a non-enveloped virus, which means it does not have a protective outer membrane. MLV replicates in the cytoplasm of infected bee cells. It is transmitted horizontally from bee to bee through contact with contaminated food, water, or hive equipment.

Transmission

MLV is primarily transmitted through contact with contaminated food or water. Infected bees can shed the virus through their faeces or saliva, which can contaminate food and water sources for other bees. MLV can also be transmitted through contact with contaminated hive equipment, such as frames, combs and honey pots.

Symptoms

The symptoms of MLV infection in honeybees can vary depending on the strain of virus involved and the severity of the infection. Some of the most common symptoms include:

- Paralysis: MLV can cause paralysis of the legs, wings and abdomen.
- Tremors: MLV can cause tremors in the legs, wings and abdomen.
- Hair loss: Infected bees may lose hair.
- Darkened abdomens: The abdomens of infected bees may darken.
- Death: MLV can cause death in infected bees.

How to identify MLV

MLV can be identified by laboratory testing of honeybee samples. The most common methods of detection include:

- Reverse transcription-polymerase chain reaction (RT-PCR): This test can detect the presence of MLV RNA in honeybee samples.
- Ensyme-linked immunosorbent assay (ELISA): This test can detect the presence of MLV antibodies in honeybee samples.

Impact on colonies

MLV infection can have a significant impact on honeybee colonies. Heavy infection can lead to:

▸ Reduced honey production: Infected colonies may produce less honey.

▸ Increased mortality: Infected colonies may experience increased mortality.

▸ Colony collapse: In severe cases, infected colonies may collapse completely.

Advice on mitigation

There is no cure for MLV infection in honeybees. However, there are a number of things that beekeepers can do to mitigate the impact of the virus:

▸ Maintain strong colony health: Strong colonies are better able to withstand the effects of MLV infection.

▸ Provide adequate nutrition: Ensure that the colony has access to a plentiful supply of pollen and other essential nutrients.

▸ Control hive temperature: Avoid extreme temperatures in the hive by providing adequate ventilation and insulation.

▸ Avoid pesticide exposure: Use pesticides only as a last resort and follow label directions carefully.

▸ Monitor colony health: Regularly inspect the colony for signs of MLV infection, such as paralysis, tremors, hair loss, darkened abdomens and death.

▸ Isolate infected colonies: If you find MLV infection in a colony, isolate it from other colonies to prevent the spread of the virus.

▸ Destroy infected hives: If a colony is heavily infected with MLV, it is best to destroy the hive and all its contents.

▸ Purchase bees from reputable beekeepers: Purchase bees from reputable beekeepers who have taken steps to control MLV in their colonies.

62. MUCOR SPP.

Mucor spp. are a group of fast growing fungi that are classified as follows:

Classification

- Kingdom: Fungi
- Phylum: Mucormycota
- Class: Mucoromycetes
- Order: Mucorales
- Family: Mucoraceae
- Genus: Mucor

Biology

Mucor spp. are fast-growing fungi that are found in soil, decaying organic matter and air. They are typically white or grey in colour and they produce distinctive sporangia (spore cases) that can be seen with the naked eye.

Mucor spp. are not typically harmful to honeybees, but they can become pathogenic if the hive is stressed or weakened. For example, Mucor spp. can cause a disease called mucormycosis in honeybees that can be fatal.

Transmission

Mucor spp. are transmitted to honeybees through contact with contaminated hive equipment, food, or water. The spores of Mucor spp. can also be carried on the bodies of other bees or insects.

Symptoms

The symptoms of mucormycosis in honeybees include:

- Lethargy
- Weakness
- Paralysis
- Death

In some cases, infected bees may also have black lesions on their bodies.

How to identify Mucor spp.

Mucor spp. can be identified by their distinctive morphology. The fungi produce white or grey mould with irregular, non-septate hyphae branching at wide angles (>90°). The sporangia of Mucor spp. are typically globose and are supported by a column-shaped columella.

Mucor spp. can also be identified using a variety of laboratory methods, such as culturing the fungus on a petri dish or using a PCR test.

Impact on colonies

Mucormycosis can have a significant impact on honeybee colonies. Heavy infestation can lead to the collapse of the colony. Mucormycosis can also weaken the colony and make it more susceptible to other diseases and pests.

Advice on mitigation

There is no cure for mucormycosis in honeybees. However, there are a number of things that beekeepers can do to mitigate the impact of the disease:

▸ Keep the hive strong and healthy.
▸ Inspect the hive regularly for signs of disease.
▸ Avoid overcrowding the hive.
▸ Use clean and sanitised hive equipment.
▸ Avoid introducing new bees into the hive without first inspecting them for disease.
▸ Purchase bees from reputable beekeepers.

If you suspect that your hive may be infected with Mucor spp., it is important to contact a beekeeper for assistance.

Additional advice on mitigation

Researchers are working to develop new ways to control mucormycosis in honeybees. One promising approach is to develop antifungal treatments that are effective against Mucor spp. Another promising approach is to develop vaccines against Mucor spp.

63. NOROVIRUS MELLIFERA

See Calicivirus

64. NOSEMA APIS

Nosema apis is a fungal like single cell organism (parasite) that is classified as follows

Classification

- Kingdom: Fungi
- Phylum: Microsporidia
- Class: Incertae sedis C
- Order: Incertae sedis O
- Family: Incertae sedis F
- Genus: Nosema
- Species: Nosema apis

Biology

Nosema Apis
Image courtesy Leila Pulsifer BC
Honey Producers Association

Nosema apis is a small, single-celled organism that is about 1-2 micrometers in diameter. The parasite replicates inside the epithelial cells of the midgut of honeybees. The replication cycle of Nosema apis is about 10-12 days long.

When a bee ingests a spore of Nosema apis, the spore infects an epithelial cell in the midgut. The parasite then replicates inside the cell, producing thousands of new spores. The spores are eventually released into the bee's faeces, where they can contaminate food and water sources and infect other bees.

Transmission

Nosema apis is transmitted from bee to bee through contact with contaminated food, water, or faeces. The spores of Nosema apis can also be transmitted on the bodies of other bees or insects.

Symptoms

The symptoms of Nosema apis infection in honeybees include:

- Reduced honey production
- Increased mortality
- Shortened lifespan

- ▶ Weakness
- ▶ Lethargy
- ▶ Diarrhoea
- ▶ Swollen abdomens
- ▶ Tremors

How to identify Nosema apis

Nosema apis can be identified by testing honeybee samples for the presence of the parasite's spores. This can be done using a variety of methods, such as microscopic examination of faeces, PCR (polymerase chain reaction), or ELISA (ensyme-linked immunosorbent assay). Nosema apis can also be identified using a compound microscope set at 400x when the spores appear as grain of rice. The parasitic spores can be identified as being smaller than nosema ceranae spores

Impact on colonies

Nosema apis can have a significant impact on honeybee colonies. Heavy infestation can lead to the collapse of the colony. Nosema apis infection can also weaken the colony and make it more susceptible to other diseases and pests.

Advice on mitigation

There is no cure for Nosema apis infection in honeybees. However, there are a number of things that beekeepers can do to mitigate the impact of the disease:

- ▶ Keep the hive strong and healthy.
- ▶ Inspect the hive regularly for signs of disease.
- ▶ Avoid overcrowding the hive.
- ▶ Use clean and sanitised hive equipment.
- ▶ Avoid introducing new bees into the hive without first inspecting them for disease.
- ▶ Purchase bees from reputable beekeepers.
- ▶ Consider using fumagillin, an antibiotic that can be used to treat Nosema apis infection.

Additional advice on mitigation

Researchers are working to develop new ways to control Nosema apis infection in honeybees. One promising approach is to develop vaccines against the parasite. Another promising approach is to use RNA interference (RNAi) to silence the genes that are responsible for the parasite's virulence.

It is important to note that Nosema apis is a complex parasite and there is still much that is not known about it.

65. NOSEMA CERANAE

Nosema ceranae is a fungal like single cell organism (parasite) that is classified as follows:

Classification

▸ Kingdom: Fungi
▸ Phylum: Microsporidia
▸ Class: Incertae sedis C
▸ Order: Incertae sedis O
▸ Family: Incertae sedis F
▸ Genus: Nosema
▸ Species: Nosema ceranae

Biology

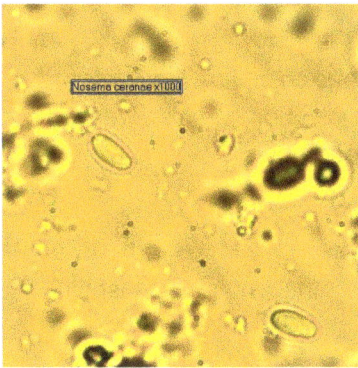

Nosema apis
Nosema apis under a microscope at X1000

Courtesy the Animal and Plant Health Agency © Crown Copyright

Nosema ceranae is a small, single-celled organism that is about 1-2 micrometers in diameter. The parasite replicates inside the epithelial cells of the midgut of honeybees. The replication cycle of Nosema ceranae is about 7-9 days long.

When a bee ingests a spore of Nosema ceranae, the spore infects an epithelial cell in the midgut. The parasite then replicates inside the cell, producing thousands of new spores. The spores are eventually released into the bee's faeces, where they can contaminate food and water sources and infect other bees.

Transmission

Nosema ceranae is transmitted from bee to bee through contact with contaminated food, water, or faeces. The spores of Nosema ceranae can also be transmitted on the bodies of other bees or insects.

Symptoms

The symptoms of Nosema ceranae infection in honeybees can be similar to those of Nosema apis infection and include:

▸ Reduced honey production
▸ Increased mortality
▸ Shortened lifespan
▸ Weakness

- Lethargy
- Diarrhoea
- Swollen abdomens
- Tremors

However, Nosema ceranae infection is generally considered to be more severe than Nosema apis infection.

How to identify Nosema ceranae

Nosema ceranae can be identified by testing honeybee samples for the presence of the parasite's spores. This can be done using a variety of methods, such as microscopic examination of faeces, PCR (polymerase chain reaction), or ELISA (ensyme-linked immunosorbent assay). Nosema ceranae can also be identified using a compound microscope set at 400x when the spores appear as grain of rice. The parasitic spores can be identified as being larger than nosema apis spores

Impact on colonies

Nosema ceranae can have a significant impact on honeybee colonies. Heavy infestation can lead to the collapse of the colony. Nosema ceranae infection can also weaken the colony and make it more susceptible to other diseases and pests.

Advice on mitigation

There is no cure for Nosema ceranae infection in honeybees. However, there are a number of things that beekeepers can do to mitigate the impact of the disease:

- Keep the hive strong and healthy.
- Inspect the hive regularly for signs of disease.
- Avoid overcrowding the hive.
- Use clean and sanitised hive equipment.
- Avoid introducing new bees into the hive without first inspecting them for disease.
- Purchase bees from reputable beekeepers.

Additional advice on mitigation

Researchers are working to develop new ways to control Nosema ceranae infection in honeybees. One promising approach is to develop vaccines against the parasite. Another promising approach is to use RNA interference (RNAi) to silence the genes that are responsible for the parasite's virulence.

It is important to note that Nosema ceranae is a complex parasite and there is still much that is not known about it.

66. ORTHOMYXOVIRUS (THOGOTOVIRUS)

Orthomyxovirus is an enveloped negative-stranded RNA virus which is classified as follows

Classification

- ▶ Kingdom: Viruses
- ▶ Realm: Riboviridae
- ▶ Phylum: Negarnaviricota
- ▶ Class: Polysomaviricetes
- ▶ Order: Negevirales
- ▶ Family: Orthomyxoviridae
- ▶ Genus: Thogotovirus

Biology

Thogotoviruses are enveloped, negative-stranded RNA viruses that replicate in the cytoplasm of infected cells. They have a wide host range, including humans, animals and insects. Honeybees are susceptible to several types of thogotoviruses, including Honeybee thogotovirus (HBThV).

HBThV replicates in the cytoplasm of infected honeybee cells, causing damage to the cells and eventually killing them. The virus is transmitted horizontally from bee to bee through contact with contaminated food, water, or hive equipment. It can also be transmitted by Varroa mites, which are parasitic mites that feed on honeybees.

Transmission

HBThV is primarily transmitted through contact with contaminated food or water. Infected adult bees can shed the virus through their faeces or saliva, which can contaminate food and water sources for other bees. HBThV can also be transmitted through contact with contaminated hive equipment, such as frames, combs and honey pots. Varroa mites can also transmit HBThV to honeybee larvae by feeding on infected larvae and then transmitting the virus to healthy larvae.

Symptoms

The symptoms of thogotovirus infection in honeybees can vary depending on the strain of virus involved and the severity of the infection. Some of the most common symptoms include:

- ▶ Reduced honey production
- ▶ Increased mortality
- ▶ Deformed wings
- ▶ Paralysis
- ▶ Hair loss
- ▶ Darkened abdomens

▸ Sticky larvae

Identification

▸ Thogotoviruses can be identified by laboratory testing of honeybee samples. The most common methods of detection include:
▸ Electron microscopy: This method can visualise the virus particles in infected larvae.
▸ Polymerase chain reaction (PCR): This test can detect the presence of thogotovirus RNA in honeybee samples.
▸ Ensyme-linked immunosorbent assay (ELISA): This test can detect the presence of thogotovirus antibodies in honeybee samples.

Impact on colonies

Thogotovirus infection can have a significant impact on honeybee colonies. Heavy infection can lead to:
▸ Reduced honey production
▸ Increased mortality
▸ Colony collapse

Advice on mitigation

There is no cure for thogotovirus infection in honeybees. However, there are a number of things that beekeepers can do to mitigate the impact of the virus:
▸ Maintain strong colony health: Strong colonies are better able to withstand the effects of thogotovirus infection.
▸ Provide adequate nutrition: Ensure that the colony has access to a plentiful supply of pollen and other essential nutrients.
▸ Control hive temperature: Avoid extreme temperatures in the hive by providing adequate ventilation and insulation.
▸ Avoid pesticide exposure: Use pesticides only as a last resort and follow label directions carefully.
▸ Monitor colony health: Regularly inspect the colony for signs of thogotovirus infection, such as reduced honey production, increased mortality, deformed wings, paralysis, hair loss, darkened abdomens and sticky larvae.
▸ Isolate infected colonies: If you find thogotovirus infection in a colony, isolate it from other colonies to prevent the spread of the virus.
▸ Destroy infected hives: If a colony is heavily infected with thogotoviruses, it is best to destroy the hive and all its contents.
▸ Purchase bees from reputable beekeepers: Purchase bees from reputable beekeepers who have taken steps to control thogotoviruses in their colonies.
▸ Monitor Varroa mite infestation: Control Varroa mite infestation as Varroa mites can transmit thogotoviruses.

67. PAENIBACILLUS ALVEI

Paenibacillus alvei is frequently found in honeybee colonies that have been weakened by EFB. It acts as a secondary invader, meaning it takes advantage of the already compromised larvae but doesn't initiate the disease itself. The presence of P. alvei in EFB-affected colonies can sometimes lead to misdiagnosis or confusion about the exact cause of the disease.

68. PAENIBACILLUS LARVAE

Paenibacillus larvae is a Gram-positive, spore-forming bacterium that is classified as follows:

Classification

- Kingdom: Bacteria
- Phylum: Firmicutes
- Class: Bacilli
- Order: Bacillales
- Family: Paenibacillaceae
- Genus: Paenibacillus
- Species: Paenibacillus larvae

Biology

Paenibacillus larvae is a rod-shaped bacterium that is about 1-2 micrometers in diameter. The bacterium is non-motile and can produce endospores, which are resistant to heat, desiccation and radiation. Paenibacillus larvae is a strict anaerobe, which means that it can only grow in the absence of oxygen.

Paenibacillus larvae is a highly specialised pathogen that only infects honeybee larvae. The spores of the bacterium are ingested by the larvae and they germinate in the gut. The vegetative cells of the bacterium then multiply rapidly and produce toxins that kill the larvae.

Transmission

Paenibacillus larvae is transmitted from bee to bee through contact with contaminated food, water, or faeces. The spores of Paenibacillus larvae can also be transmitted on the bodies of other bees or insects.

Symptoms

The symptoms of American foulbrood (AFB), the disease caused by Paenibacillus larvae, include:

- Deformed larvae
- Discoloured larvae
- Stunted larvae
- Dead larvae
- A sour odour from the hive

How to identify Paenibacillus larvae

Paenibacillus larvae can be identified by culturing the bacterium from honeybee samples or by using PCR (polymerase chain reaction) to test for the presence of the bacterium's DNA.

Impact on colonies

AFB can have a devastating impact on honeybee colonies. Heavy infestation can lead to the collapse of the colony. AFB can also weaken the colony and make it more susceptible to other diseases and pests.

Advice on mitigation

There is no cure for AFB. However, there are a number of things that beekeepers can do to mitigate the impact of the disease:

- ▶ Keep the hive strong and healthy.
- ▶ Inspect the hive regularly for signs of disease.
- ▶ Avoid overcrowding the hive.
- ▶ Use clean and sanitised hive equipment.
- ▶ Avoid introducing new bees into the hive without first inspecting them for disease.
- ▶ Purchase bees from reputable beekeepers.
- ▶ Requeen the colony if it is heavily infested with Paenibacillus larvae.

Additional advice on mitigation

Researchers are working to develop new ways to control AFB. One promising approach is to develop vaccines against Paenibacillus larvae. Another promising approach is to use bacteriophages, which are viruses that specifically infect bacteria, to kill Paenibacillus larvae.

It is important to note that Paenibacillus larvae is a complex bacterium and there is still much that is not known about it..

69. PENICILLIUM SPP.

Penicillium spp. is a diverse group of fungi which are classified as follows

Classification

▶ Kingdom: Fungi
▶ Phylum: Ascomycota
▶ Class: Eurotiomycetes
▶ Order: Eurotiales
▶ Family: Trichomaceae
▶ Genus: Penicillium

Penicillium spp. are a diverse group of fungi that can be found in a variety of environments, including soil, water and plants. There are over 1,000 known species of Penicillium and some of these species can be pathogenic to honeybees.

Biology

Penicillium spp. are saprophytic fungi, meaning that they get their nutrients from dead and decaying matter. They can also be facultative parasites, meaning that they can live on both dead and living matter. When Penicillium spp. infect a honeybee, they enter the bee through the mouth or spiracles (breathing holes) and grow in the bee's hemolymph (blood). The fungus produces toxins that damage the bee's cells and organs. The infection can spread throughout the body, eventually leading to the death of the bee.

Transmission

Penicillium spp. are transmitted from bee to bee through contact with contaminated faeces or saliva. Infected adult bees can shed the fungus through their faeces or saliva, which can contaminate food and water sources for other bees. Penicillium spp. can also be transmitted by Varroa mites, which are parasitic mites that feed on honeybees.

Symptoms

The symptoms of Penicillium spp. infection can vary depending on the severity of the infection and the age of the bee. Some of the most common symptoms include:

▶ Reduced honey production
▶ Increased mortality
▶ Deformed wings
▶ Paralysis
▶ Hair loss
▶ Darkened abdomens
▶ Sticky larvae
▶ Swollen abdomens
▶ Discoloured pupae
▶ Stunted growth
▶ Poor flying ability

Identification

Penicillium spp. can be identified by laboratory testing of honeybee samples. The most common methods of detection include:

▶ Culture: This method can grow the fungus in a laboratory setting, which allows for further identification.

▶ Polymerase chain reaction (PCR): This test can detect the presence of Penicillium spp. DNA in honeybee samples.

▶ Ensyme-linked immunosorbent assay (ELISA): This test can detect the presence of Penicillium spp. antibodies in honeybee samples.

Impact on Colonies

Penicillium spp. can have a significant impact on honeybee colonies. Heavy infection can lead to:

▶ Reduced honey production

▶ Increased mortality

▶ Colony collapse

Advice on Mitigation

There is no cure for Penicillium spp. infection in honeybees. However, there are a number of things that beekeepers can do to mitigate the impact of the fungus. These include:

▶ Maintain strong colony health: Strong colonies are better able to withstand the effects of Penicillium spp.

▶ Provide adequate nutrition: Ensure that the colony has access to a plentiful supply of pollen and other essential nutrients.

▶ Control hive temperature: Avoid extreme temperatures in the hive by providing adequate ventilation and insulation.

▶ Avoid pesticide exposure: Use pesticides only as a last resort and follow label directions carefully.

▶ Monitor colony health: Regularly inspect the colony for signs of Penicillium spp. infection, such as reduced honey production, increased mortality, deformed wings, paralysis, hair loss, darkened abdomens, sticky larvae, swollen abdomens and discoloured pupae.

▶ Isolate infected colonies: If you find Penicillium spp. infection in a colony, isolate it from other colonies to prevent the spread of the fungus.

▶ Destroy infected hives: If a colony is heavily infected with Penicillium spp., it is best to destroy the hive and all its contents.

▶ Purchase bees from reputable beekeepers: Purchase bees from reputable beekeepers who have taken steps to control Penicillium spp. infection in their colonies.

▶ Monitor Varroa mite infestation: Control Varroa mite infestation as Varroa mites can transmit the fungus.Purchase bees from reputable beekeepers: Purchase bees from reputable beekeepers who have taken steps to control Penicillium spp. infection in their colonies.

70. POLYOMAVIRUS

Polyomavirus is a non-enveloped double-stranded DNA virus which is classified as follows

Classification

- ▶ Kingdom: Viruses
- ▶ Realm: Riboviridae
- ▶ Phylum: Negarnaviricota
- ▶ Class: DNA
- ▶ Order: Polyomavirales
- ▶ Family: Polyomaviridae

Biology

Polyomaviruses are non-enveloped, double-stranded DNA viruses that replicate in the nucleus of infected cells. They have a wide host range, including humans, animals and birds. Honeybees are susceptible to several types of polyomaviruses, including Honeybee polyomavirus (HBPyV).

HBPyV replicates in the nucleus of infected honeybee cells, causing damage to the cells and eventually killing them. The virus is transmitted horizontally from bee to bee through contact with contaminated food, water, or hive equipment.

Transmission

HBPyV is primarily transmitted through contact with contaminated food or water. Infected adult bees can shed the virus through their faeces or saliva, which can contaminate food and water sources for other bees. HBPyV can also be transmitted through contact with contaminated hive equipment, such as frames, combs and honey pots.

Symptoms

The symptoms of polyomavirus infection in honeybees can vary depending on the strain of virus involved and the severity of the infection. Some of the most common symptoms include:

- ▶ Reduced honey production
- ▶ Increased mortality
- ▶ Deformed wings
- ▶ Paralysis
- ▶ Hair loss
- ▶ Darkened abdomens

Identification

Polyomaviruses can be identified by laboratory testing of honeybee samples. The most common methods of detection include:

▸ Electron microscopy: This method can visualise the virus particles in infected larvae.

▸ Polymerase chain reaction (PCR): This test can detect the presence of polyomavirus DNA in honeybee samples.

▸ Ensyme-linked immunosorbent assay (ELISA): This test can detect the presence of polyomavirus antibodies in honeybee samples.

Impact on colonies

Polyomavirus infection can have a significant impact on honeybee colonies. Heavy infection can lead to:

▸ Reduced honey production

▸ Increased mortality

▸ Colony collapse

Advice on mitigation

There is no cure for polyomavirus infection in honeybees. However, there are a number of things that beekeepers can do to mitigate the impact of the virus:

▸ Maintain strong colony health: Strong colonies are better able to withstand the effects of polyomavirus infection.

▸ Provide adequate nutrition: Ensure that the colony has access to a plentiful supply of pollen and other essential nutrients.

▸ Control hive temperature: Avoid extreme temperatures in the hive by providing adequate ventilation and insulation.

▸ Avoid pesticide exposure: Use pesticides only as a last resort and follow label directions carefully.

▸ Monitor colony health: Regularly inspect the colony for signs of polyomavirus infection, such as reduced honey production, increased mortality, deformed wings, paralysis, hair loss and darkened abdomens.

▸ Isolate infected colonies: If you find polyomavirus infection in a colony, isolate it from other colonies to prevent the spread of the virus.

▸ Destroy infected hives: If a colony is heavily infected with polyomaviruses, it is best to destroy the hive and all its contents.

▸ Purchase bees from reputable beekeepers: Purchase bees from reputable beekeepers who have taken steps to control polyomaviruses in their colonies.

71. PROTEUS MIRABILIS

Proteus mirabilis and honeybees have a complex relationship. Here's a breakdown:

Classification

- Kingdom: Bacteria
- Phylum: Proteobacteria
- Class: Gammaproteobacteria
- Order: Enterobacterales
- Family: Enterobacteriaceae
- Species: Proteus mirabilis

Biology

Proteus mirabilis is a Gram-negative, rod-shaped bacterium known for its swarming motility and ability to produce urease, an enzyme that breaks down urea. It's a common inhabitant of soil, water, and even the human gut (where it usually resides harmlessly).

Transmission

The presence of Proteus mirabilis in honeybees is still being explored. Here are two possibilities:

- Commensal Gut Bacteria: Proteus mirabilis might be part of the honeybee's natural gut microbiome, existing in small numbers without causing harm. Some studies suggest it might even offer some protection against other pathogens [1].
- Environmental Contamination: Proteus mirabilis could enter the hive through contaminated food sources (pollen, nectar) or water. This is more likely to cause problems.

Symptoms

Unfortunately, there aren't specific symptoms for Proteus mirabilis infection in honeybees. However, a stressed or weakened hive with high overall bacterial load might be more susceptible to opportunistic infections, including those by Proteus mirabilis.

Identification

Identifying Proteus mirabilis requires specialised laboratory testing to isolate and culture the bacteria.

Impact on Colonies

While some studies suggest a protective role, high levels of Proteus mirabilis can be detrimental to honey bee health. Potential problems include:

- Dysentery: Proteus mirabilis can disrupt the gut microbiome, leading to diarrhea and dysentery in bees.
- Weak Immune System: A compromised gut can weaken the bee's immune system, making it more susceptible to other diseases.

▸ Reduced Honey Production: A sick hive is a less productive hive.

Mitigation Strategies

Since the role of Proteus mirabilis in honeybees is still under investigation, definitive mitigation strategies are limited. However, general beekeeping practices that promote colony health can help:

▸ Strong Genetics: Healthy bee breeds with good resistance to diseases are a good starting point.

▸ Nutrition: Provide high-quality pollen substitutes and ensure access to clean water sources.

▸ Hive Management: Regular hive inspections and proper hive hygiene can help prevent the spread of pathogens.

▸ Minimising Stress: Reduce stress factors for bees such as pesticide use and excessive manipulation.

Future Research

More research is needed to understand the exact role of Proteus mirabilis in honeybee health. Studies on its potential as a commensal gut bacterium and its impact on pathogen resistance are ongoing.

72. PSEUDOMONAS AERUGINOSA

Pseudomonas aeruginosa is a gram-negative, rod shaped bacterium which is classified as follows

Classification

▶ Kingdom: Bacteria
▶ Phylum: Proteobacteria
▶ Class: Gammaproteobacteria
▶ Order: Pseudomonadales
▶ Family: Pseudomonadaceae
▶ Genus: Pseudomonas
▶ Species: P. aeruginosa

Biology

Pseudomonas aeruginosa is a Gram-negative, rod-shaped bacterium that is commonly found in soil, water and plants. It is an opportunistic pathogen that can cause a wide range of infections in humans, animals and insects. Honeybees are susceptible to P. aeruginosa, which can cause a disease called Pseudomonas apiseptica.

P. aeruginosa replicates in the hemolymph (blood) of infected honeybees, causing damage to the bees' cells and eventually killing them. The bacterium is transmitted horizontally from bee to bee through contact with contaminated food, water, or hive equipment. It can also be transmitted by Varroa mites, which are parasitic mites that feed on honeybees.

Transmission

P. aeruginosa is primarily transmitted through contact with contaminated food or water. Infected adult bees can shed the bacterium through their faeces or saliva, which can contaminate food and water sources for other bees. P. aeruginosa can also be transmitted through contact with contaminated hive equipment, such as frames, combs and honey pots. Varroa mites can also transmit P. aeruginosa to honeybee larvae by feeding on infected larvae and then transmitting the bacterium to healthy larvae.

Symptoms

The symptoms of P. aeruginosa infection in honeybees can vary depending on the strain of bacterium involved and the severity of the infection. Some of the most common symptoms include:

▶ Reduced honey production
▶ Increased mortality
▶ Deformed wings
▶ Paralysis
▶ Hair loss
▶ Darkened abdomens

- Sticky larvae
- Swollen abdomens
- Discoloured pupae

Identification

Pseudomonas aeruginosa infection can have a significant impact on honeybee colonies. Heavy infection can lead to:
- Reduced honey production
- Increased mortality
- Colony collapse

Advice on Mitigation

There is no cure for Pseudomonas aeruginosa infection in honeybees. However, there are a number of things that beekeepers can do to mitigate the impact of the bacterium:

- Maintain strong colony health: Strong colonies are better able to withstand the effects of P. aeruginosa infection.
- Provide adequate nutrition: Ensure that the colony has access to a plentiful supply of pollen and other essential nutrients.
- Control hive temperature: Avoid extreme temperatures in the hive by providing adequate ventilation and insulation.
- Avoid pesticide exposure: Use pesticides only as a last resort and follow label directions carefully.
- Monitor colony health: Regularly inspect the colony for signs of P. aeruginosa infection, such as reduced honey production, increased mortality, deformed wings, paralysis, hair loss, darkened abdomens, sticky larvae, swollen abdomens and discoloured pupae.
- Isolate infected colonies: If you find P. aeruginosa infection in a colony, isolate it from other colonies to prevent the spread of the bacterium.
- Destroy infected hives: If a colony is heavily infected with P. aeruginosa, it is best to destroy the hive and all its contents.
- Purchase bees from reputable beekeepers: Purchase bees from reputable beekeepers who have taken steps to control P. aeruginosa in their colonies.
- Monitor Varroa mite infestation: Control Varroa mite infestation as Varroa mites can transmit P. aeruginosa.

73. REOVIRUS

Reorvirus is a non-enveloped double-stranded RNA virus which is classified as follows

Classification

- ▶ Kingdom: Viruses
- ▶ Realm: Riboviridae
- ▶ Phylum: Negarnaviricota
- ▶ Class: Duplviricota
- ▶ Order: Reovirales
- ▶ Family: Reoviridae
- ▶ Genus: Orbivirus

Biology

Reoviridae are non-enveloped, double-stranded RNA viruses that replicate in the cytoplasm of infected cells. They have a wide host range, including humans, animals and insects.

Reoviruses replicate in the cytoplasm of infected honeybee cells, causing damage to the cells and eventually killing them. The viruses are transmitted horizontally from bee to bee through contact with contaminated food, water, or hive equipment. They can also be transmitted by Varroa mites, which are parasitic mites that feed on honeybees.

Transmission

Reoviruses are primarily transmitted through contact with contaminated food or water. Infected adult bees can shed the viruses through their faeces or saliva, which can contaminate food and water sources for other bees. Reoviruses can also be transmitted through contact with contaminated hive equipment, such as frames, combs and honey pots. Varroa mites can also transmit reoviruses to honeybee larvae by feeding on infected larvae and then transmitting the viruses to healthy larvae.

Symptoms

The symptoms of reovirus infection in honeybees can vary depending on the strain of virus involved and the severity of the infection. Some of the most common symptoms include:

- ▶ Reduced honey production
- ▶ Increased mortality
- ▶ Deformed wings
- ▶ Paralysis
- ▶ Hair loss
- ▶ Darkened abdomens
- ▶ Sticky larvae

Identification

Reoviridae can be identified by laboratory testing of honeybee samples. The most common methods of detection include:

▶ Electron microscopy: This method can visualise the virus particles in infected larvae.

▶ Polymerase chain reaction (PCR): This test can detect the presence of reovirus RNA in honeybee samples.

▶ Ensyme-linked immunosorbent assay (ELISA): This test can detect the presence of reovirus antibodies in honeybee samples.

Impact on colonies

Reovirus infection can have a significant impact on honeybee colonies. Heavy infection can lead to:

▶ Reduced honey production

▶ Increased mortality

▶ Colony collapse

Advice on mitigation

There is no cure for reovirus infection in honeybees. However, there are several things that beekeepers can do to mitigate the impact of the viruses:

▶ Maintain strong colony health: Strong colonies are better able to withstand the effects of reovirus infection.

▶ Provide adequate nutrition: Ensure that the colony has access to a plentiful supply of pollen and other essential nutrients.

▶ Control hive temperature: Avoid extreme temperatures in the hive by providing adequate ventilation and insulation.

▶ Avoid pesticide exposure: Use pesticides only as a last resort and follow label directions carefully.

▶ Monitor colony health: Regularly inspect the colony for signs of reovirus infection, such as reduced honey production, increased mortality, deformed wings, paralysis, hair loss, darkened abdomens and sticky larvae.

▶ Isolate infected colonies: If you find reovirus infection in a colony, isolate it from other colonies to prevent the spread of the virus.

▶ Destroy infected hives: If a colony is heavily infected with reoviruses, it is best to destroy the hive and all its contents.

▶ Purchase bees from reputable beekeepers: Purchase bees from reputable beekeepers who have taken steps to control reoviruses in their colonies.

▶ Monitor Varroa mite infestation: Control Varroa mite infestation as Varroa mites can transmit reoviruses.

74. RHABDOVIRUS

Rhabdoviruses pose a growing threat to honeybee health. Here's a detailed breakdown of their biology, classification, transmission, symptoms, identification, impact on colonies, and mitigation strategies:

Classification

Honeybee rhabdoviruses can be classified into two categories:

- Classified:
  - Family: Rhabdoviridae
  - Genus: Nucleorhabdovirus
    - Species: Arkansas bee virus (ABV) - A significant pathogen associated with colony decline.
- Unclassified:
  - Bee rhabdovirus-1 (BRV-1) & BRV-2: Their role in bee health is unclear and their full classification is under investigation.
  - Other potential rhabdoviruses: Recent research suggests a wider diversity of rhabdoviruses within honeybee populations.

Biology

Rhabdoviruses are negative-sense, single-stranded RNA viruses. Their defining feature is a bullet-shaped structure with a protein envelope surrounding the RNA core. They replicate inside host cells, producing new viral particles that can infect other cells.

Transmission

The primary mode of transmission for rhabdoviruses in honeybees is through the parasitic mite, Varroa destructor.

- Mite Acquisition: Varroa mites pick up the virus while feeding on infected bees.
- Viral Replication: The virus replicates within the mite.
- Transmission to Healthy Bees: During subsequent feeding, the mite transmits the virus to healthy bees.

Other potential transmission routes are being investigated:

- Horizontal transmission: Direct contact between bees.
- Vertical transmission: From queen bee to larvae.

Symptoms

Unfortunately, rhabdovirus infections in honeybees often lack clear external signs. However, some potential indicators include:

- Reduced bee activity: Fewer bees foraging or visible around the hive.
- Deformed wings: Particularly in developing bees.
- Increased bee mortality: Higher than usual bee deaths.

Identification

Diagnosing rhabdovirus infection requires laboratory techniques:

▸ RT-PCR (Reverse Transcription Polymerase Chain Reaction): This detects the viral genetic material in samples collected from bees.

▸ Viral isolation: Attempts to grow the virus in cell cultures for further analysis.

Impact on Colonies

Rhabdoviruses, can have a devastating impact on honeybee colonies:

▸ Increased colony mortality: Infected colonies are more likely to collapse.

▸ Reduced bee lifespan: The virus can shorten the lifespan of individual bees.

▸ Weakened immune system: Bees infected with rhabdoviruses may be more susceptible to other diseases.

▸ Reduced honey production: Colony decline due to rhabdovirus can affect honey yields.

Mitigation Strategies

While there's no current cure for rhabdovirus infections, beekeepers can take steps to minimise their impact:

▸ Integrated Pest Management (IPM) for Varroa mites: Regular monitoring and control of Varroa mite populations is crucial.

▸ Strong colony management: Healthy colonies with good genetics are better equipped to fight off infections.

▸ Nutritional support: Providing essential nutrients can help strengthen bee immune systems.

▸ Emerging Research: Ongoing research may lead to future vaccines or treatments for rhabdoviruses.

Additional Notes

▸ The full extent of rhabdovirus diversity and their exact role in honeybee health is still being explored.

▸ Early detection and intervention are critical for managing rhabdovirus infections.

▸ Consulting with a local beekeeping association or veterinarian can provide valuable guidance for beekeepers concerned about rhabdoviruses.

By understanding rhabdoviruses and taking preventative measures, beekeepers can help protect their honeybee colonies and ensure the health of these vital pollinators.

75. RHISOPUS SPP

Rhisopus spp. is a group of fast growing fungi that are classified as follows:

Classification

▸ Kingdom: Fungi
▸ Phylum: Mucormycota
▸ Class: Mucoromycetes
▸ Order: Mucorales
▸ Family: Mucoraceae
▸ Genus: Rhisopus

Biology

Rhisopus spp. are fast-growing fungi that are found in soil, decaying organic matter and air. They are typically white or grey in colour and they produce distinctive sporangia (spore cases) that can be seen with the naked eye.

Rhisopus spp. are not typically harmful to honeybees, but they can become pathogenic if the hive is stressed or weakened. For example, Rhisopus spp. can cause a disease called mucormycosis in honeybees that can be fatal.

Transmission

Rhisopus spp. are transmitted to honeybees through contact with contaminated hive equipment, food, or water. The spores of Rhisopus spp. can also be carried on the bodies of other bees or insects.

Symptoms

The symptoms of mucormycosis in honeybees include:

▸ Lethargy
▸ Weakness
▸ Paralysis
▸ Death

In some cases, infected bees may also have black lesions on their bodies.

How to identify Rhisopus spp.

Rhisopus spp. can be identified by their distinctive morphology. The fungi produce white or grey mold with irregular, non-septate hyphae branching at wide angles (>90°). The sporangia of Rhisopus spp. are typically globose and are supported by a column-shaped columella.

Rhisopus spp. can also be identified using a variety of laboratory methods, such as culturing the fungus on a petri dish or using a PCR (polymerase chain reaction) test.

Impact on colonies

Mucormycosis can have a significant impact on honeybee colonies. Heavy infestation can lead to the collapse of the colony. Mucormycosis can also weaken the colony and make it more susceptible to other diseases and pests.

Advice on mitigation

There is no cure for mucormycosis in honeybees. However, there are several things that beekeepers can do to mitigate the impact of the disease:

- Keep the hive strong and healthy.
- Inspect the hive regularly for signs of disease.
- Avoid overcrowding the hive.
- Use clean and sanitised hive equipment.
- Avoid introducing new bees into the hive without first inspecting them for disease.
- Purchase bees from reputable beekeepers.
- If you suspect that your hive may be infected with Rhisopus spp., it is important to contact a beekeeper for assistance.

Additional advice on mitigation

Researchers are working to develop new ways to control mucormycosis in honeybees. One promising approach is to develop antifungal treatments that are effective against Rhisopus spp. Another promising approach is to develop vaccines against Rhisopus spp.

It is important to note that Rhisopus spp. are a diverse group of fungi and there is still much that is not known about them.

76. SAC BROOD VIRUS

Sacbrood is a viral disease that affects honeybee larvae. The virus responsible for sacbrood is called sacbrood virus (SBV), which is classified as follows:

Classification

- ▶ Kingdom: Viruses
- ▶ Phylum: Riboviria
- ▶ Order: Picornavirales
- ▶ Family: Iflaviridae
- ▶ Genus: Ilfavirus
- ▶ Species: Sacbrood virus

Biology

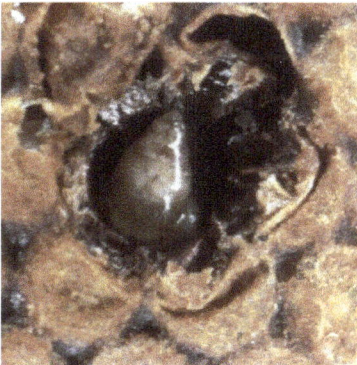

Sacbrood Virus
Sacbrood infected larvae forms a fluid filled sack that is easily removed from the cell. Its shape is said to resemble a Chinese Slip

Courtesy the Animal and Plant Health Agency © Crown Copyright

Sacbrood virus is a single-stranded RNA virus that replicates in the cytoplasm of honeybee larvae. The virus enters the larvae through their gut and then replicates rapidly, eventually killing the larvae. The dead larvae become filled with fluid, giving them a sac-like appearance, which is how the disease gets its name.

Transmission

Sacbrood virus is transmitted from bee to bee through contact with contaminated food, water, or faeces. The virus can also be transmitted through contaminated hive equipment.

Symptoms

The symptoms of sacbrood include:

- ▶ Discoloured larvae
- ▶ Sunken or perforated cappings
- ▶ Dead larvae with a sac-like appearance
- ▶ A sour odour from the hive

Infected larvae typically die within 3-5 days of being infected.

How to identify sacbrood

Sacbrood can be identified by examining the brood pattern in the hive. Infected larvae will typically be discoloured and have sunken or perforated cappings. The dead larvae will have a sac-like appearance and may be scattered throughout the hive.

Sacbrood can also be diagnosed by laboratory testing of honeybee samples. This can be done using a variety of methods, such as PCR (polymerase chain reaction) or ELISA (ensyme-linked immunosorbent assay).

Impact on colonies

Sacbrood can have a significant impact on honeybee colonies. Heavy infestation can lead to the collapse of the colony. Sacbrood can also weaken the colony and make it more susceptible to other diseases and pests.

Sacbrood
Sacbrood infected larva that has been removed from comb. The dead larva resembles a fluid filled sack

Courtesy the Animal and Plant Health Agency © Crown Copyright

Advice on mitigation

There is no cure for sacbrood. However, there are several things that beekeepers can do to mitigate the impact of the disease:

▸ Keep the hive strong and healthy.
▸ Inspect the hive regularly for signs of disease.
▸ Avoid overcrowding the hive.
▸ Use clean and sanitised hive equipment.
▸ Avoid introducing new bees into the hive without first inspecting them for disease.
▸ Purchase bees from reputable beekeepers.
▸ Requeen the colony if it is heavily infested with sacbrood virus..

Additional advice on mitigation

Researchers are working to develop new ways to control sacbrood. One promising approach is to develop vaccines against sacbrood virus. Another promising approach is to use RNA interference (RNAi) to silence the genes that are responsible for the virus's virulence.

It is important to note that sacbrood virus is a complex virus and there is still much that is not known about it.

77. SEOUL BEE VIRUS

Seoul bee virus (SBV) is a single-stranded RNA virus that belongs to the family Picornaviridae. It is classified as follows:

Classification

- Kingdom: Viruses
- Phylum: Riboviria
- Class: Picornavirales
- Order: Dicistroviridae
- Family: Dicistroviridae
- Genus: Iridovirus
- Species: Seoul bee virus (SBV)

Biology

SBV is a relatively small virus, with a genome that is about 10,000 nucleotides long. The virus replicates in the cytoplasm of honeybee cells. The first step in the replication cycle is the attachment of the virus to the cell surface. The virus then enters the cell and its RNA is translated into proteins. The proteins form a complex that replicates the viral RNA. The new viral RNA is then translated into proteins, which form the new virus particles.

Transmission

SBV is transmitted from bee to bee through contact with contaminated food, water, or faeces. The virus can also be transmitted through contaminated hive equipment.

Symptoms

The symptoms of SBV infection in honeybees are not well-defined. However, some studies have shown that SBV infection can be associated with:

- Reduced queen acceptance
- Increased mortality of bees
- Weakened colony

How to identify SBV

SBV infection can be identified by testing honeybee samples for the presence of the virus's RNA. This can be done using a variety of methods, such as real-time PCR (qPCR) or reverse transcription PCR (RT-PCR).

Impact on colonies

The impact of SBV infection on honeybee colonies is not fully understood. However, some studies have shown that SBV infection can be associated with:

- ▶ Reduced queen acceptance
- ▶ Increased mortality of bees
- ▶ Weakened colony
- ▶ Collapse of the colony

Advice on mitigation

There is no cure for SBV infection in honeybees. However, there are several things that beekeepers can do to mitigate the impact of the disease:

- ▶ Keep the hive strong and healthy.
- ▶ Inspect the hive regularly for signs of disease.
- ▶ Avoid overcrowding the hive.
- ▶ Use clean and sanitised hive equipment.
- ▶ Avoid introducing new bees into the hive without first inspecting them for disease.
- ▶ Purchase bees from reputable beekeepers.

Additional advice on mitigation

Researchers are working to develop new ways to control SBV infection. One promising approach is to develop vaccines against the virus. Another promising approach is to use RNA interference (RNAi) to silence the genes that are responsible for the virus's virulence.

It is important to note that SBV is a complex virus and there is still much that is not known about it.

78. SERRATIA MARCESCENS

Serratia marcescens is a Gram-negative, rod-shaped bacterium that is classified as follows:

Classification

- Kingdom: Bacteria
- Phylum: Firmicutes
- Class: Bacilli
- Order: Pseudomonadales
- Family: Enterobacteriaceae
- Genus: Serratia
- Species: Serratia marcescens

Biology

Serratia marcescens is a ubiquitous bacterium that is found in soil, water, plants and animals. It is an opportunistic pathogen, meaning that it can cause infection in people and honeybees with weakened immune systems. Serratia marcescens is also known for its ability to produce a red pigment called prodigiosin, which can give it a red or pink colour.

Transmission

Serratia marcescens is transmitted to honeybees through contact with contaminated food, water, or faeces. The bacteria can also be transmitted through contaminated hive equipment.

Symptoms

The symptoms of Serratia marcescens infection in honeybees include:
- Discoloured larvae
- Stunted larvae
- Dead larvae
- A sour odour from the hive

How to identify Serratia marcescens

Serratia marcescens can be identified by culturing the bacterium from honeybee samples or by using PCR (polymerase chain reaction) to test for the presence of the bacterium's DNA.

Impact on colonies

Serratia marcescens infection can have a significant impact on honeybee colonies. Heavy infestation can lead to the collapse of the colony. Serratia marcescens infection can also weaken the colony and make it more susceptible to other diseases and pests.

Advice on mitigation

There is no cure for Serratia marcescens infection in honeybees. However, there are several things that beekeepers can do to mitigate the impact of the disease:

▸ Keep the hive strong and healthy.
▸ Inspect the hive regularly for signs of disease.
▸ Avoid overcrowding the hive.
▸ Use clean and sanitised hive equipment.
▸ Avoid introducing new bees into the hive without first inspecting them for disease.
▸ Purchase bees from reputable beekeepers.

Additional advice on mitigation

Researchers are working to develop new ways to control Serratia marcescens infection in honeybees. One promising approach is to develop vaccines against the bacterium. Another promising approach is to use bacteriophages, which are viruses that specifically infect bacteria, to kill Serratia marcescens.

It is important to note that Serratia marcescens is a complex bacterium and there is still much that is not known about it..

79. SLOW BEE PARALYSIS VIRUS

Slow bee paralysis virus (SBPV) is a single-stranded RNA virus that belongs to the family Iflaviridae. It is classified as follows:

Classification

▶ Kingdom: Viruses
▶ Phylum: Riboviria
▶ Class: Picornavirales
▶ Order: Dicistroviridae
▶ Family: Iflaviridae
▶ Genus: Iflavirus
▶ Species: Slow bee paralysis virus (SBPV)

Biology

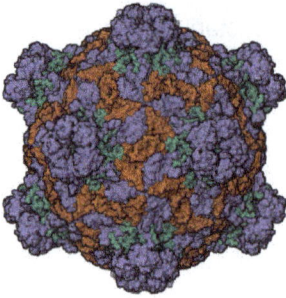

Slow Bee Paralysis Virus

Courtesy rcsb.org
5J96 Crystal structure of Slow Bee Paralysis Virus at 3.4A resolution
PDB DOI: https://doi.org/10.2210/pdb5J96/pdb

Deposition Author(s): Kalynych, S., Levdansky, Y., Palkova, L., Plevka, P.

SBPV is a relatively small virus, with a genome that is about 9.5 kb long. The virus replicates in the cytoplasm of honeybee cells. The first step in the replication cycle is the attachment of the virus to the cell surface. The virus then enters the cell and its RNA is translated into proteins. The proteins form a complex that replicates the viral RNA. The new viral RNA is then translated into proteins, which form the new virus particles.

Transmission

SBPV is transmitted from bee to bee through contact with contaminated food, water, or faeces. The virus can also be transmitted through contaminated hive equipment.

Symptoms

The symptoms of SBPV infection in honeybees include:

▶ Paralysis in the front two pairs of legs
▶ Inability to fly
▶ Tremors
▶ Death

The symptoms of SBPV infection typically develop 10-12 days after exposure to the virus. The infected bees typically die within 2-3 weeks of showing symptoms.

How to identify SBPV

SBPV infection can be identified by testing honeybee samples for the presence of the virus's RNA. This can be done using a variety of methods, such as real-time PCR (qPCR) or reverse transcription PCR (RT-PCR).

Impact on colonies

SBPV infection can have a significant impact on honeybee colonies. Heavy infestation can lead to the collapse of the colony. SBPV infection can also weaken the colony and make it more susceptible to other diseases and pests.

Advice on mitigation

There is no cure for SBPV infection in honeybees. However, there are several things that beekeepers can do to mitigate the impact of the disease:

▶ Keep the hive strong and healthy.
▶ Inspect the hive regularly for signs of disease.
▶ Avoid overcrowding the hive.
▶ Use clean and sanitised hive equipment.
▶ Avoid introducing new bees into the hive without first inspecting them for disease.
▶ Purchase bees from reputable beekeepers.

Additional advice on mitigation

Researchers are working to develop new ways to control SBPV infection. One promising approach is to develop vaccines against the virus. Another promising approach is to use RNA interference (RNAi) to silence the genes that are responsible for the virus's virulence.

It is important to note that SBPV is a complex virus and there is still much that is not known about it. However, by following the mitigation advice above, beekeepers can help to protect their colonies from the devastating effects of SBPV infection.

80. SLOW BEE PARALYSIS VIRUS-LIKE VIRUS

Slow bee paralysis virus-like virus (SBPV-LLV) is a single-stranded RNA virus that is morphologically and genetically similar to slow bee paralysis virus (SBPV). However, SBPV-LLV is not considered to be a true member of the Iflaviridae family, as it lacks a conserved sequence in its genome.

Classification

‣ Kingdom: Viruses
‣ Phylum: Riboviria
‣ Class: Picornavirales
‣ Order: Dicistroviridae
‣ Family: Unassigned

Biology

SBPV-LLV is a relatively small virus, with a genome that is about 9.5 kb long. The virus replicates in the cytoplasm of honeybee cells. The first step in the replication cycle is the attachment of the virus to the cell surface. The virus then enters the cell and its RNA is translated into proteins. The proteins form a complex that replicates the viral RNA. The new viral RNA is then translated into proteins, which form the new virus particles.

Transmission

SBPV-LLV is transmitted from bee to bee through contact with contaminated food, water, or faeces. The virus can also be transmitted through contaminated hive equipment.

Symptoms

The symptoms of SBPV-LLV infection in honeybees are similar to those of SBPV infection and include:

‣ Paralysis in the front two pairs of legs
‣ Inability to fly
‣ Tremors
‣ Death

The symptoms of SBPV-LLV infection typically develop 10-12 days after exposure to the virus. The infected bees typically die within 2-3 weeks of showing symptoms.

How to identify SBPV-LLV

SBPV-LLV infection can be identified by testing honeybee samples for the presence of the virus's RNA. This can be done using a variety of methods, such as real-time PCR (qPCR) or reverse transcription PCR (RT-PCR).

Impact on colonies

SBPV-LLV infection can have a significant impact on honeybee colonies. Heavy infestation can lead to the collapse of the colony. SBPV-LLV infection can also weaken the colony and make it more susceptible to other diseases and pests.

Advice on mitigation

There is no cure for SBPV-LLV infection in honeybees. However, there are several things that beekeepers can do to mitigate the impact of the disease:
- Keep the hive strong and healthy.
- Inspect the hive regularly for signs of disease.
- Avoid overcrowding the hive.
- Use clean and sanitised hive equipment.
- Avoid introducing new bees into the hive without first inspecting them for disease.
- Purchase bees from reputable beekeepers.

Additional advice on mitigation

Researchers are working to develop new ways to control SBPV-LLV infection. One promising approach is to develop vaccines against the virus. Another promising approach is to use RNA interference (RNAi) to silence the genes that are responsible for the virus's virulence.

It is important to note that SBPV-LLV is a complex virus and there is still much that is not known about it. However, by following the mitigation advice above, beekeepers can help to protect their colonies from the devastating effects of SBPV-LLV infection.

82. SMALL HIVE BEETLE

The small hive beetle (SHB), Aethina tumida, is a globally significant pest of honeybees, posing a considerable threat to colony health and honey production. This invasive beetle, native to sub-Saharan Africa, has spread to honeybee regions worldwide, prompting beekeepers to stay vigilant. Let's delve into the details of this formidable foe, exploring its classification, biology, transmission, symptoms, identification, impact, and mitigation strategies.

Classification:

- ▶ Domain: Eukaryota
- ▶ Kingdom: Animalia
- ▶ Phylum: Arthropoda
- ▶ Class: Insecta
- ▶ Order: Coleoptera
- ▶ Family: Histeridae
- ▶ Genus: Aethina

Adult Small Hive Beetle
Closeup view of an Adult Small Hive Beetle

Courtesy the Animal and Plant Health Agency © Crown Copyright

Biology

Small hive beetles are small, reddish-brown to black insects, measuring about 5-7mm in length. They have clubbed antennae, characteristic wrinkles on their wing covers, and strong legs for crawling and clinging. Adults can fly up to 15 km, aiding in their ability to locate and infest new hives. Both adults and larvae feed on honey, pollen, brood, and comb structure, damaging the hive and impacting bee health.

Transmission:

Flight: Adult beetles actively search for new hives and can fly long distances to infest vulnerable colonies.

Hitchhiking: SHBs can inadvertently hitch rides on beekeeping equipment, vehicles, or bee swarms.

Contaminated honey: Beehives containing brood or honey contaminated with beetle eggs can spread the infestation when moved.

Symptoms of Infestation:

Presence of adult beetles: Identifying reddish-brown beetles crawling on frames, honeycomb, or the hive floor is the most direct sign.

Damaged brood and comb: Evidence of chewed brood cells, deformed larvae, and slimy, fermented honey due to beetle feeding.

Weakened colony: Reduced bee activity, increased bee mortality, and honey production decline can indicate SHB presence.

Sour honey: Honey contaminated with beetle larvae or excrement often exhibits a sour or fermented smell and taste.

Identification:

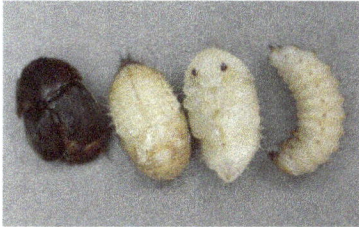

Small Hive Beetle
Small Hive Beetle (SHB) life stages
Beetle, Pupae, Larvae

Courtesy the Animal and Plant Health Agency © Crown Copyright

Visual inspection: Carefully examine frames, comb surfaces, and the hive floor for adult beetles or their characteristic larval tunnels.

Trapping: Using beetle traps placed near hive entrances can capture adults for confirmation.

Honey test: Checking for a sour odor or fermented taste in extracted honey can raise suspicion of SHB contamination.

Professional assistance: In some cases, consulting a beekeeping expert or veterinarian may be necessary for definitive identification and advice.

Impact on Colonies:

Direct damage: Feeding by adults and larvae destroys brood, honey, and comb, reducing colony resources and impacting bee development.

Weakening effect: Infested colonies become weakened, making them more susceptible to other diseases and parasites.

Honey contamination: Beetles and their larvae contaminate honey with bacteria and fungi, rendering it unfit for consumption.

Colony collapse: Severe infestations can lead to queen failure, brood decline, and eventually, colony collapse.

Mitigation Strategies:

Preventative measures:

Strong colony management: Strong, healthy colonies with good hygiene practices are better equipped to resist SHB infestations.

Hive hygiene: Maintain clean equipment, avoid introducing contaminated materials, and regularly remove debris from the hive floor.

Inspection and monitoring: Regularly inspect hives for adult beetles, larvae, or damage, and use traps for early detection.

Control measures:

Chemical treatments: Apply registered pesticides or insect growth regulators following manufacturer instructions and beekeeping regulations.

Organic methods: Diatomaceous earth, essential oils, and parasitic nematodes can offer environmentally friendly control options.

Traps: Bait traps can attract and capture adult beetles, aiding in population reduction.

Community cooperation: Implementing coordinated control measures within an apiary or wider beekeeping community can significantly limit SHB spread.

83. SOLITARY BEE VIRUS

Solitary bee viruses (SBVs) are a group of viruses that can infect solitary bees. SBVs can also infect honeybees (Apis mellifera) but they are not as common in honeybees as they are in solitary bees.

Classification

- Kingdom: Viruses
- Realm: Riboviria
- Phylum: Negarnaviricota
- Class: Picornaviricetes
- Order: Picornavirales
- Family: Dicistroviridae
- Genus: Crimoviridae

Biology

SBVs are single-stranded RNA viruses. They replicate in the cytoplasm of infected cells. SBVs are transmitted horizontally from bee to bee through contact with contaminated food, water, or hive equipment. They can also be transmitted through Varroa mites, which are parasitic mites that feed on honeybees.

Transmission

SBVs are primarily transmitted through contact with contaminated food or water. Infected bees can shed the virus through their faeces or saliva, which can contaminate food and water sources for other bees. SBVs can also be transmitted through contact with contaminated hive equipment, such as frames, combs and honey pots, but this is less common. Varroa mites can also transmit SBVs to honeybees by feeding on infected bees and then transmitting the virus to healthy bees.

Symptoms

The symptoms of SBV infection in honeybees can vary depending on the strain of virus involved and the severity of the infection. Some of the most common symptoms include:

- Paralysis: SBVs can cause paralysis of the legs, wings and abdomen.
- Tremors: SBVs can cause tremors in the legs, wings and abdomen.
- Hair loss: Infected bees may lose hair.
- Darkened abdomens: The abdomens of infected bees may darken.
- Deformed wings: SBVs can cause deformed wings in developing bees.
- Death: SBVs can cause death in infected bees.

How to identify SBV

SBVs can be identified by laboratory testing of honeybee samples. The most common methods of detection include:

▶ Reverse transcription-polymerase chain reaction (RT-PCR): This test can detect the presence of SBV RNA in honeybee samples.

▶ Ensyme-linked immunosorbent assay (ELISA): This test can detect the presence of SBV antibodies in honeybee samples.

Impact on colonies

SBV infection can have a significant impact on honeybee colonies. Heavy infection can lead to:

▶ Reduced honey production: Infected colonies may produce less honey.

▶ Increased mortality: Infected colonies may experience increased mortality.

▶ Colony collapse: In severe cases, infected colonies may collapse completely.

Advice on mitigation

There is no cure for SBV infection in honeybees. However, there are several things that beekeepers can do to mitigate the impact of the virus:

▶ Maintain strong colony health: Strong colonies are better able to withstand the effects of SBV infection.

▶ Provide adequate nutrition: Ensure that the colony has access to a plentiful supply of pollen and other essential nutrients.

▶ Control hive temperature: Avoid extreme temperatures in the hive by providing adequate ventilation and insulation.

▶ Avoid pesticide exposure: Use pesticides only as a last resort and follow label directions carefully.

▶ Monitor colony health: Regularly inspect the colony for signs of SBV infection, such as paralysis, tremors, hair loss, darkened abdomens, deformed wings and death.

▶ Isolate infected colonies: If you find SBV infection in a colony, isolate it from other colonies to prevent the spread of the virus.

▶ Destroy infected hives: If a colony is heavily infected with SBV, it is best to destroy the hive and all its contents.

▶ Purchase bees from reputable beekeepers: Purchase bees from reputable beekeepers who have taken steps to control SBV in their colonies.

▶ Monitor Varroa mite infestation: Control Varroa mite infestation as Varroa mites can transmit SBVs.

84. SPIROPLASMOSIS

Spiroplasmosis is a bacterium which lives in the hemolymph of apis mellifera and is classified as follows

Classification

- ▶ Kingdom: Bacteria
- ▶ Phylum: Firmicutes
- ▶ Class: Mollicutes
- ▶ Order: Entomoplasmatales
- ▶ Family: Spiroplasmataceae
- ▶ Genus: Spiroplasma
- ▶ Species: Spiroplasma melliferum

Biology

Spiroplasma melliferum is a bacterium that lives in the hemolymph (blood) of honeybees. It is a helical-shaped bacterium that is about 200 nanometers long. The bacterium is transmitted from bee to bee through contact with contaminated faeces or saliva.

Transmission

Spiroplasmosis is most commonly transmitted through contact with contaminated food or water. Infected adult bees can shed the bacterium through their faeces or saliva, which can contaminate food and water sources for other bees. Spiroplasmosis can also be transmitted by Varroa mites, which are parasitic mites that feed on honeybees.

Symptoms

The symptoms of spiroplasmosis can vary depending on the strain of bacteria involved and the severity of the infection. Some of the most common symptoms include:

- ▶ Reduced honey production
- ▶ Increased mortality
- ▶ Deformed wings
- ▶ Paralysis
- ▶ Hair loss
- ▶ Darkened abdomens
- ▶ Sticky larvae
- ▶ Swollen abdomens
- ▶ Discoloured pupae

Identification

Spiroplasmosis can be identified by laboratory testing of honeybee samples. The most common methods of detection include:

▸ Culture: This method can grow the bacterium in a laboratory setting, which allows for further identification.

▸ Polymerase chain reaction (PCR): This test can detect the presence of Spiroplasma melliferum DNA in honeybee samples.

▸ Ensyme-linked immunosorbent assay (ELISA): This test can detect the presence of Spiroplasma melliferum antibodies in honeybee samples.

Impact on Colonies

Spiroplasmosis can have a significant impact on honeybee colonies. Heavy infection can lead to:

▸ Reduced honey production
▸ Increased mortality
▸ Colony collapse

Advice on Mitigation

There is no cure for spiroplasmosis in honeybees. However, there are several things that beekeepers can do to mitigate the impact of the bacteria:

▸ Maintain strong colony health: Strong colonies are better able to withstand the effects of spiroplasmosis.

▸ Provide adequate nutrition: Ensure that the colony has access to a plentiful supply of pollen and other essential nutrients.

▸ Control hive temperature: Avoid extreme temperatures in the hive by providing adequate ventilation and insulation.

▸ Avoid pesticide exposure: Use pesticides only as a last resort and follow label directions carefully.

▸ Monitor colony health: Regularly inspect the colony for signs of spiroplasmosis, such as reduced honey production, increased mortality, deformed wings, paralysis, hair loss, darkened abdomens, sticky larvae, swollen abdomens and discoloured pupae.

▸ Isolate infected colonies: If you find spiroplasmosis in a colony, isolate it from other colonies to prevent the spread of the bacteria.

▸ Destroy infected hives: If a colony is heavily infected with spiroplasmosis, it is best to destroy the hive and all its contents.

▸ Purchase bees from reputable beekeepers: Purchase bees from reputable beekeepers who have taken steps to control spiroplasmosis in their colonies.

▸ Monitor Varroa mite infestation: Control Varroa mite infestation as Varroa mites can transmit bacteria.

85. STREPTOCOCCUS FAECALIS

Classification Streptococcus faecalis is a Gram-positive, spherical bacterium that belongs to the family Streptococcaceae. It is classified as follows:

Classification

- ▸ Kingdom: Bacteria
- ▸ Phylum: Firmicutes
- ▸ Class: Bacilli
- ▸ Order: Lactobacillales
- ▸ Family: Streptococcaceae
- ▸ Genus: Streptococcus
- ▸ Species: Streptococcus faecalis

Biology

Streptococcus faecalis is a facultative anaerobe, meaning that it can grow with or without oxygen. It is a relatively hardy bacterium that can survive in a variety of environments, including soil, water and food. Streptococcus faecalis is also a common inhabitant of the human gut microbiome.

Transmission

Streptococcus faecalis is transmitted to honeybees through contact with contaminated food, water, or faeces. The bacteria can also be transmitted through contaminated hive equipment.

Symptoms

Streptococcus faecalis infection in honeybees can cause a variety of symptoms, including:

- ▸ Diarrhoeal
- ▸ Dysentery
- ▸ Poor growth and development
- ▸ Increased mortality

How to identify Streptococcus faecalis Streptococcus faecalis can be identified by culturing the bacterium from honeybee samples or by using PCR (polymerase chain reaction) to test for the presence of the bacterium's DNA.

Impact on colonies

Streptococcus faecalis infection can have a significant impact on honeybee colonies. Heavy infestation can lead to the collapse of the colony. Streptococcus faecalis infection can also weaken the colony and make it more susceptible to other diseases and pests.

Advice on mitigation

There is no cure for Streptococcus faecalis infection in honeybees. However, there are several things that beekeepers can do to mitigate the impact of the disease:

▸ Keep the hive strong and healthy.
▸ Inspect the hive regularly for signs of disease.
▸ Avoid overcrowding the hive.
▸ Use clean and sanitised hive equipment.
▸ Avoid introducing new bees into the hive without first inspecting them for disease.
▸ Purchase bees from reputable beekeepers.
▸ Additional advice on mitigation
▸ Researchers are working to develop new ways to control Streptococcus faecalis infection in honeybees. One promising approach is to develop vaccines against the bacterium. Another promising approach is to use bacteriophages, which are viruses that specifically infect bacteria, to kill Streptococcus faecalis.
▸ It is important to note that Streptococcus faecalis is a complex bacterium and there is still much that is not known about it.

Additional advice on mitigation

Researchers are working to develop new ways to control Streptococcus faecalis infection in honeybees. One promising approach is to develop vaccines against the bacterium. Another promising approach is to use bacteriophages, which are viruses that specifically infect bacteria, to kill Streptococcus faecalis.

It is important to note that Streptococcus faecalis is a complex bacterium and there is still much that is not known about it.

86. THOGOTOVIRUS

See Orthomyxoviridae

87. TOBACCO RINGSPOT VIRUS

Tobacco ringspot virus (TRSV) is a single-stranded RNA virus that belongs to the family Secoviridae. It is classified as follows:

Classification

- ▶ Kingdom: Viruses
- ▶ Phylum: Riboviria
- ▶ Class: Nepovirus
- ▶ Order: Mononegavirales
- ▶ Family: Secoviridae
- ▶ Genus: Nepovirus
- ▶ Species: Tobacco ringspot virus (TRSV)

Biology

TRSV is a relatively small virus, with a genome that is about 6 kb long. The virus replicates in the cytoplasm of honeybee cells. The first step in the replication cycle is the attachment of the virus to the cell surface. The virus then enters the cell and its RNA is translated into proteins. The proteins form a complex that replicates the viral RNA. The new viral RNA is then translated into proteins, which form the new virus particles.

Transmission

TRSV is transmitted to honeybees through contact with contaminated pollen or sap. The virus can also be transmitted through contaminated hive equipment.

Symptoms

The symptoms of TRSV infection in honeybees are not well-defined. However, some studies have shown that TRSV infection can be associated with:

- ▶ Reduced queen acceptance
- ▶ Increased mortality of bees
- ▶ Weakened colony

How to identify TRSV

TRSV infection can be identified by testing honeybee samples for the presence of the virus's RNA. This can be done using a variety of methods, such as real-time PCR (qPCR) or reverse transcription PCR (RT-PCR).

Impact on colonies

The impact of TRSV infection on honeybee colonies is not fully understood. However, some studies have shown that TRSV infection can be associated with:

- Reduced queen acceptance
- Increased mortality of bees
- Weakened colony
- Collapse of the colony

Advice on mitigation

There is no cure for TRSV infection in honeybees. However, there are several things that beekeepers can do to mitigate the impact of the disease:

- Keep the hive strong and healthy.
- Inspect the hive regularly for signs of disease.
- Avoid overcrowding the hive.
- Use clean and sanitised hive equipment.
- Avoid introducing new bees into the hive without first inspecting them for disease.
- Purchase bees from reputable beekeepers.

Additional advice on mitigation

Researchers are working to develop new ways to control TRSV infection. One promising approach is to develop vaccines against the virus. Another promising approach is to use RNA interference (RNAi) to silence the genes that are responsible for the virus's virulence.

It is important to note that TRSV is a complex virus and there is still much that is not known about it. However, by following the mitigation advice above, beekeepers can help to protect their colonies from the devastating effects of TRSV infection.

88. TOGAVIRUS

Togavirus is an enveloped, positive-stranded RNA virus which is classified as follows

Classification

- Kingdom: Viruses
- Realm: Riboviridae
- Phylum: Negarnaviricota
- Class: Alphavirus
- Order: Togaviriles
- Family: Togaviridae
- Genus: Alphavirus

Biology

Togaviridae are enveloped, positive-stranded RNA viruses that replicate in the cytoplasm of infected cells. They have a wide host range, including humans, animals and insects. Honeybees are susceptible to several types of togaviruses.

Togaviruses replicate in the cytoplasm of infected honeybee larvae, causing damage to the cells and eventually killing them. The viruses are transmitted horizontally from bee to bee through contact with contaminated food, water, or hive equipment..

Transmission

Togaviruses are primarily transmitted through contact with contaminated food or water. Infected adult bees can shed the viruses through their faeces or saliva, which can contaminate food and water sources for other bees. They can also be transmitted through contact with contaminated hive equipment, such as frames, combs and honey pots.

Symptoms

The symptoms of togavirus infection in honeybees can vary depending on the strain of virus involved and the severity of the infection. Some of the most common symptoms include:

- Reduced honey production
- Increased mortality
- Deformed wings
- Paralysis
- Hair loss
- Darkened abdomens
- Sticky larvae

Identification

- Togaviridae can be identified by laboratory testing of honeybee samples. The most common methods of detection include:

▶ Electron microscopy: This method can visualise the virus particles in infected larvae.

▶ Polymerase chain reaction (PCR): This test can detect the presence of togavirus RNA in honeybee samples.

▶ Ensyme-linked immunosorbent assay (ELISA): This test can detect the presence of togavirus antibodies in honeybee samples.

Impact on colonies

Togavirus infection can have a significant impact on honeybee colonies. Heavy infection can lead to:

▶ Reduced honey production
▶ Increased mortality
▶ Colony collapse

Advice on mitigation

There is no cure for togavirus infection in honeybees. However, there are several things that beekeepers can do to mitigate the impact of the viruses:

▶ Maintain strong colony health: Strong colonies are better able to withstand the effects of togavirus infection.

▶ Provide adequate nutrition: Ensure that the colony has access to a plentiful supply of pollen and other essential nutrients.

▶ Control hive temperature: Avoid extreme temperatures in the hive by providing adequate ventilation and insulation.

▶ Avoid pesticide exposure: Use pesticides only as a last resort and follow label directions carefully.

▶ Monitor colony health: Regularly inspect the colony for signs of togavirus infection, such as reduced honey production, increased mortality, deformed wings, paralysis, hair loss, darkened abdomens and sticky larvae.

▶ Isolate infected colonies: If you find togavirus infection in a colony, isolate it from other colonies to prevent the spread of the virus.

▶ Destroy infected hives: If a colony is heavily infected with togaviruses, it is best to destroy the hive and all its contents.

▶ Purchase bees from reputable beekeepers: Purchase bees from reputable beekeepers who have taken steps to control togaviruses in their colonies.

▶ Monitor Varroa mite infestation: Control Varroa mite infestation as Varroa mites can transmit togaviruses

89. TROPILAELAPS CLAREAE

Tropilaelaps clareae: A Devastating Parasite of honeybees

Classification:

▸ Kingdom: Animalia
▸ Phylum: Arthropoda
▸ Class: Arachnida
▸ Subclass: Acari
▸ Superorder: Parasitiformes
▸ Order: Mesostigmata
▸ Family: Laelapidae
▸ Species: Tropilaelaps clareae

Biology

Tropilaelaps mite Alongside Varroa Mite
Close up view of Tropilaelaps mite and Varroa Mite

Courtesy the Animal and Plant Health Agency © Crown Copyright

Life cycle:
▸ Egg-laying: Female mites lay 1-4 eggs within capped bee pupae.
▸ Larval stage: Six-legged larvae hatch and feed on the developing bee pupae.
▸ Nymphal stage: Larvae molt twice into eight-legged nymphs, still feeding on pupae.
▸ Adult stage: Nymphs molt into adult mites (1 mm long, reddish-brown), which mate and reproduce.
▸ Lifecycle length: ~10-14 days under optimal conditions.

Transmission

▸ Honeybee drift: Individual bees moving between hives.
▸ Robbing: Attacking and stealing honey from other colonies.
▸ Swarming: Infested bees migrating to form new colonies.
▸ Beekeeping practices: Sharing brood combs, moving colonies, buying/selling infected bees.

Symptoms of Tropilaelaps clareae Infestation:

▸ Deformed or dead pupae: Parasitism by mites stunts and disrupts development.
▸ Reduced brood production: Fewer healthy bees emerge due to pupal mortality and weakened colonies.
▸ Weakened colony: Smaller population, sluggish behaviour, increased susceptibility to other diseases and predators.
▸ Perforated cappings: Worker bees may remove infested pupae, leaving holes in

the capped cells.
▶ Crawling bees: Some infested bees exhibit disorientation and crawl near the hive entrance.

Tropilaelaps on brood
Tropilaelaps mites parasitising honeybee pupa

Courtesy the Animal and Plant Health Agency © Crown Copyright

Identification:

▶ Visual inspection: Look for adult mites on brood combs (fast-moving) or reddish-brown eggs inside capped cells.
▶ Microscopic examination: Dissected dead pupae may reveal mite larvae and nymphs.

Impact on Colonies:

▶ Reduced honey production: Fewer bees directly impact honey collection.
▶ Colony decline: Infestation weakens and shrinks colonies, increasing vulnerability to other threats.
▶ Increased mortality: Severe infestations can lead to colony collapse.

Mitigation Strategies:

▶ Early detection: Regular hive inspections to identify infestations early.
▶ Chemical control: Acaricides specifically targeting Tropilaelaps mites (resistance is a growing concern).
▶ Cultural control: Practices like queen replacement, brood removal, and hive splitting.
▶ Integrated pest management (IPM): Combining multiple control methods for long-term efficacy

90. TROPILAELAPS MERCEDESAE

Tropilaelaps mercedesae: A Growing Menace to Honeybees

Classification:

- ▶ Kingdom: Animalia
- ▶ Phylum: Arthropoda
- ▶ Class: Arachnida
- ▶ Subclass: Acari
- ▶ Superorder: Parasitiformes
- ▶ Order: Mesostigmata
- ▶ Family: Laelapidae
- ▶ Species: Tropilaelaps mercedesae

Biology

- ▶ Life cycle: Similar to T. clareae, with 1-4 eggs laid in pupae, followed by larval and nymphal stages before reaching adulthood.
- ▶ Lifecycle length: Slightly longer than T. clareae, around 14-18 days.

Transmission:

Same as T. clareae, including honeybee drift, robbing, swarming, and beekeeping practices.

Symptoms of Tropilaelaps mercedesae Infestation:

- ▶ Deformed or dead pupae: Common to both Tropilaelaps species.
- ▶ Reduced brood production: More pronounced with T. mercedesae, with larvae and pupae being attacked.
- ▶ Weak and disoriented bees: Emerging bees can suffer developmental issues due to parasite feeding.
- ▶ Increased crawling bees: More frequent than with T. clareae, indicating neurological damage.
- ▶ Dysfunctional behaviours: Abnormal hive activities like erratic egg-laying and aggression.
- ▶ Perforated cappings and sanitation attempts: Frequent due to higher mortality rates.

Identification:

- ▶ Visual inspection: Similar to T. clareae, look for reddish-brown mites on combs or eggs within capped cells.
- ▶ Microscopic examination: Remains the most reliable method to differentiate T. mercedesae from T. clareae.

Impact on Colonies:

▶ Severe brood damage: T. mercedesae targets both larvae and pupae, causing higher mortality and developmental deformities.

▶ Colony decline and potential collapse: Faster and more devastating decline compared to T. clareae infestations.

▶ Impaired foraging and defense: Weakened and disoriented bees struggle to perform vital tasks.

▶ Increased susceptibility to other diseases: Stress from infestation makes colonies vulnerable to secondary infections.

Mitigation Strategies:

▶ Early detection and swift action: Regular inspections are crucial due to the rapid colony decline.

▶ Chemical control: Targeted acaricides can be effective, but resistance is a concern.

▶ Integrated pest management (IPM): Combining methods like brood removal, queen replacement, and chemical control for long-term control.

▶ Quarantine procedures: In high-risk areas, isolate new colonies and monitor for infestations.

91. VARROA DESTRUCTOR

The mite Varroa destructor casts a long shadow over the world of honeybees. Understanding its classification, biology, transmission, and impact is crucial for protecting these vital pollinators. So, let's delve into the realm of this parasitic villain:

Classification:

- Kingdom: Animalia
- Phylum: Arthropoda
- Subphylum: Hexapoda
- Class: Arachnida
- Subclass: Acari
- Order: Mesostigmata
- Family: Varroidae
- Genus: Varroa
- Species: Varroa destructor

Effects of varroa on honey bees
Example of deformity of bees from Varroosis

Courtesy the Animal and Plant Health Agency © Crown Copyright

Varroa Mite
Close up of a Varroa Mite on comb

Courtesy the Animal and Plant Health Agency © Crown Copyright

Biology:

- Size: About the size of a pinhead, reddish-brown in colour.
- Lifecycle: Females are primarily reproductive, laying eggs inside capped honeybee pupae. They feed on the fat body of developing bees, weakening them and transmitting viruses. Males fertilize females within the brood cells before dying.
- Reproduction: Parthenogenetic (females can reproduce without males) in some generations, increasing their reproductive potential.

Transmission:

▶ Primary: Varroa mites primarily crawl and hitchhike between adult bees during brood rearing and social interactions.

▶ Secondary: Transmission can also occur when drones mate with virgin queens or through robbing behaviour between colonies.

Symptoms:

▶ Deformed wings: A classic symptom, but not always present. Occurs when larvae are infected, hindering their development.

▶ Weakened bees: Sluggish movement, trembling, and difficulty flying might indicate mite infestation.

▶ Increased mortality: Higher death rates, particularly among drones and young bees, can be a telltale sign.

▶ Colony weakening: Reduced brood production, smaller honey stores, and increased susceptibility to other diseases are consequences of a heavy infestation.

Varroa infestation
Severe Varroa infestation with young bee covered in mites and showing signs of deformity
Courtesy the Animal and Plant Health Agency © Crown Copyright

Varroa damage
Adult bees and brood suffering deformities, poor health and death caused by a significant Varroa infestation
Courtesy the Animal and Plant Health Agency © Crown Copyright

Identification:

▶ Visual inspection: Experienced beekeepers can spot mites on adult bees, although this method is not always reliable.

▶ Sugar shake: Dusting bees with powdered sugar dislodges mites, making them easier to see.

▶ Alcohol wash: Immersing a frame of bees in alcohol dissolves dead mites, revealing their number.

▶ Varroa mite counts: Regularly monitoring mite populations through these methods is crucial for timely intervention.

Impact on colonies:

▶ Transmission of viruses: Varroa mites are primary vectors for deformed wing virus and other bee viruses, significantly impacting colony health.

▶ Weakened immune system: Constant feeding depletes bees' energy and compromises their immune function, making them vulnerable to other diseases and pathogens.

▶ Reduced productivity: Decreased brood production, honey yields, and pollination services are a direct consequence of colony weakening.

Mitigation:

▶ Integrated pest management (IPM): Utilizing a combination of methods is key, including:
- Chemical control: Mite-specific acaricides, applied with careful consideration and proper timing.
- Organic approaches: Essential oils, brood manipulation techniques, and resistant bee breeds can offer sustainable options.
- Monitoring and record-keeping: Regularly checking mite levels and adapting treatment strategies based on data is essential.

▶ Maintaining strong colonies: Good beekeeping practices like providing adequate nutrition, proper hive hygiene, and stress reduction promote colony health and resilience against mites.

Remember: Varroa destructor is a major threat to honeybees, but knowledge and proactive management can help protect these vital pollinators. By understanding its biology, transmission, and impact, and implementing effective mitigation strategies, beekeepers can safeguard their colonies and contribute to a healthy bee population.

100. VARROA JACOBSONI

While Varroa destructor grabs the headlines as the primary honeybee mite pest, its less-talked-about cousin, Varroa jacobsoni, also deserves our attention. Let's shed light on this enigmatic parasite:

Classification:

- Kingdom: Animalia
- Phylum: Arthropoda
- Subphylum: Hexapoda
- Class: Arachnida
- Subclass: Acari
- Order: Mesostigmata
- Family: Varroidae
- Genus: Varroa
- Species: Varroa jacobsoni

Biology:

Size is similar to V. destructor, reddish-brown in colour. Its lifecycle is similar to V. destructor, with females breeding inside capped bee pupae and males dying after fertilisation.

Reproduction is primarily parthenogenetic, capable of reproducing without males, further enhancing its potential spread and its host preference is historically parasitised Asian honeybees (Apis cerana), but recent evidence suggests adaptation to European honeybees (Apis mellifera), raising concerns about its potential impact.

Transmission:

Similar to V. destructor: Primarily through direct contact between bees during brood rearing and social interactions.Research suggests potential transmission through drone mating flights and robbing behaviour, requiring further investigation.

Symptoms in Honeybees:

Due to its recent adaptation to Apis mellifera, definitive symptoms specific to V. jacobsoni infestations are still unclear. Potential overlap with V. destructor include deformed wings, weakened bees, increased mortality, and colony weakening might occur, but differentiation from V. destructor requires specific testing.

More studies are needed to establish distinct symptom profiles and diagnostic methods for V. jacobsoni infestations in Apis mellifera.

Identification:

- Microscopic examination: Currently, the only reliable method for distinguishing V. jacobsoni from V. destructor involves detailed analysis of their mouthparts and leg morphology under a microscope.

▶ DNA testing: PCR testing can accurately identify the mite species, but might not be readily accessible to all beekeepers.

Impact on Colonies:

The full extent of V. jacobsoni's impact on Apis mellifera colonies remains unclear, but its documented role in transmitting viruses and weakening colonies in Apis cerana raises concerns. If V. jacobsoni becomes fully established and spreads widely in Apis mellifera populations, it could contribute to their decline, adding to the existing challenges faced by these critical pollinators.

Mitigation:

▶ Focus on Varroa destructor control: As V. destructor remains the more widespread and detrimental threat, effective control measures against it are crucial. This includes integrated pest management practices like chemical treatments, organic approaches, and monitoring mite populations.

▶ Enhanced surveillance: Close monitoring of bee health and potential V. jacobsoni infestations, particularly in regions where it has been detected, is vital for early detection and mitigation efforts.

▶ Research and development: Continued research on V. jacobsoni's biology, transmission, impact, and potential control strategies is essential for safeguarding bee health and preventing its expansion.

Remember: While the threat posed by V. jacobsoni to Apis mellifera needs careful monitoring and further investigation, vigilance against V. destructor through established control measures remains paramount. By taking proactive steps and supporting ongoing research, we can protect our beloved honeybees from both these parasitic shadows.

92. VARROA DESTRUCTOR VIRUS-1

Varroa destructor virus type 1 (VDV!) has more recently been described as being a variation of deformed wing virus type B (DWV-B) and has now lost its classification.

Refer to deformed wing virus type B

93. VARROA DESTRUCTOR VIRUS-2

Although this book primarily is looking at honeybee diseases, it is valid to include varroa destructor virus type 2 (VDV-2). This virus impacts the varroa mite and has a reported impact on reducing the varroa mite population. However, it is also reported that varroa mites which are infected with this virus, also have a greater impact when it comes to transmitting DWV A/B/C to the colony. No decisive assessment has yet been concluded.

VDV-2, a single-stranded RNA virus shrouded in intrigue, inhabits the microscopic battleground within Varroa destructor mites, parasitic adversaries of honeybees. Let's dive deep into its world, exploring its classification, biology, transmission, and potential impact on both mites and their honeybee hosts.

Classification:

▶ Domain: Viruses
▶ Kingdom: Riboviria
▶ Phylum: Pisuviricota
▶ Class: Pisoniviricetes
▶ Order: Picornavirales
▶ Family: Iflaviridae
▶ Genus: Iflavirus
▶ Species: Iflavirus vadestruente

Biology:

Genome: Composed of three segments of single-stranded RNA, encoding proteins crucial for replication and assembly.

Replication: Occurs within the mite's cytoplasm, involving the production of viral RNA and protein components.

Transmission

Primarily horizontal, occurring during mite-to-mite contact. Vertical transmission (mother to offspring) is also possible.

Symptoms in Varroa Mites:

Currently, no distinct symptoms are attributed to VDV-2 infection in Varroa mites. They appear to tolerate the virus well, with no obvious physical or behavioural changes.

Impact on Colonies:

The complex relationship between VDV-2 and honeybee health is an ongoing research puzzle.

Studies suggest VDV-2 may not directly harm bees but can potentially influence the severity of other bee viruses, especially Deformed wing virus (DWV).

Possible ways VDV-2 impacts colonies:

 Increased DWV transmission: Varroa mites infected with VDV-2 appear to transmit DWV to bees more efficiently.

Immunosuppression: VDV-2 might weaken the bee's immune system, making them more susceptible to other viruses.

Indirect effects: Increased viral infections and weakened immunity can lead to colony decline, reduced honey production, and increased susceptibility to other pathogens.

Current Research:

Scientists are actively investigating the precise mechanisms by which VDV-2 interacts with DWV and other bee viruses.

Understanding these interactions is crucial for developing effective control measures against Varroa mites and the viruses they carry, ultimately protecting honeybee health and ensuring their critical role in pollination.

Intriguing Questions

Does VDV-2 possess any evolutionary advantage for the Varroa mite?

Can specific strains of VDV-2 influence its impact on bee health?

Can understanding VDV-2 lead to novel mite control strategies that minimise harmful effects on bees?

A Tapestry of Complexity

VDV-2 remains a fascinating enigma, interwoven with the intricate web of mite-bee interactions and viral dynamics. Its classification provides a crucial framework, but its true role in the honeybee health saga is still being written. Continued research promises to unveil new chapters, offering insights that could safeguard these vital pollinators and the ecosystems they sustain.

94. WAX MOTH

See Bald Brood

Wax moth larvae
Wax moth larvae in webbing causing damage in an empty beehive

Courtesy the Animal and Plant
Health Agency © Crown Copyright

95. WEIL'S DISEASE

Weil's disease, also known as Leptospirosis, is a bacterial infection that can affect honeybees. It is caused by the bacteria Leptospira interrogans, which is commonly found in water and soil. Bees can become infected by contact with contaminated water, soil, or other contaminated materials.

Classification

- Kingdom: Bacteria
- Phylum: Spirochaetes
- Genus: Leptospira
- Species: interrogans

Biology

Leptospira interrogans is a spiral-shaped bacterium that can survive in a variety of environmental conditions. It can be found in the urine of infected animals, such as rodents and livestock. The bacteria can also survive in water and soil for long periods of time.

Transmission

Bees can become infected with Weil's disease through contact with contaminated water, soil, or other contaminated materials. When a bee comes into contact with the bacteria, the bacteria can enter the bee's body through the mouth or skin. The bacteria can then spread through the bee's body and cause infection.

Symptoms of Weil's Disease

The symptoms of Weil's disease in honeybees can vary depending on the strain of the bacteria, the age of the bee, and the overall health of the colony. Some of the most common symptoms include:

▶ Reduced brood viability: Weil's disease can infect honeybee larvae and cause them to die or emerge with developmental abnormalities.

▶ Impaired brood development: Weil's disease can disrupt the normal development of honeybee larvae, leading to delayed pupation and emergence time.

▶ Reduced adult bee performance: Weil's disease-infected adult bees may appear lethargic, have reduced foraging activity, and exhibit impaired learning and memory.

▶ Increased susceptibility to other diseases: Weil's disease infection can weaken the immune system of honeybees, making them more susceptible to other pathogens such as Varroa destructor and Nosema ceranae.

Identifying Weil's Disease

There are no specific clinical tests available for the detection of Weil's disease in individual honeybees. However, the presence of the bacteria can be inferred based on the observation of characteristic symptoms and the detection of the bacteria in hive samples.

Symptoms: Monitoring colonies for symptoms such as reduced brood viability, impaired brood development, and poor adult bee performance can suggest Weil's disease infection.

Hive samples: Laboratory analysis of hive samples, such as dead bees, pollen, and brood, can detect the presence of the bacteria.

Impact on Colonies

Weil's disease can have a significant impact on honeybee colonies, contributing to colony decline. The bacteria can disrupt the development of larvae, leading to reduced brood viability and colony population. Additionally, infected adult bees may exhibit impaired foraging ability, further impacting the colony's productivity.

Mitigation Strategies

There is no specific treatment or cure for Weil's disease in honeybees. However, several preventive and management strategies can be implemented to reduce the risk of outbreaks and minimise the impact of the bacterium:

▶ Strong Colony Management: Maintaining healthy, thriving colonies with adequate nutrition, ventilation, and hygienic practices can boost the overall resistance of bees to infections.

▶ Monitoring and Control of Varroa Mites: Varroa mites are major vectors of bacterial infections, including Weil's disease. Effective Varroa mite control is crucial for overall colony health.

▶ Protecting Hives from Pesticides: Pesticide exposure can disrupt the gut microbiome of honeybees and reduce their resistance to diseases. Avoiding or minimising pesticide use around hives can help protect bee health.

▶ Providing Adequate Space: Overcrowding can stress honeybees and make them more susceptible to infections. Providing ample space for colonies to grow and thrive can help reduce the risk of outbreaks.

▶ By implementing these strategies, beekeepers can contribute to the prevention and management of Weil's disease outbreaks and promote the overall health of honeybee colonies

ACKNOWLEDGEMENTS

Firstly I acknowledge with thanks the support of Jeremy Burbige, the publisher, who has been outstanding in his encouragement to develop an AI led book on honeybee diseases.

I would like to thank Kirsty Stainton for her invaluable support at a technical review level, which has been so helpful when wading through the various AI outputs.

I would also like to thank John Phipps who completed the initial review and David Miller who has the patience of a saint.

Finally, I wish to acknowledge the various sources of images used in this book, including the Animal and Plant Health Agency (APHA) Rachel Green of the State University of New Jersey, Leilani Pulsifer of the British Columbia Honey Producers Association and Autumn Canaday from the USDA-Agricultural Research Service.

INDEX   PAGE

www.ingramcontent.com/pod-product-compliance
Lightning Source LLC
Chambersburg PA
CBHW080546220326
41599CB00032B/6383